Springer Texts in Business and Economics

Springer Texts in Business and Economics (STBE) delivers high-quality instructional content for undergraduates and graduates in all areas of Business/Management Science and Economics. The series is comprised of self-contained books with a broad and comprehensive coverage that are suitable for class as well as for individual self-study. All texts are authored by established experts in their fields and offer a solid methodological background, often accompanied by problems and exercises.

Dobromir Kirilov Stoyanov

Sustainable Marketing

A Value-based Approach

Springer

Dobromir Kirilov Stoyanov
EM-Strasbourg Business School, HuManiS
Research Center (UR 7308)
University of Strasbourg
Strasbourg Cedex, France

ISSN 2192-4333 ISSN 2192-4341 (electronic)
Springer Texts in Business and Economics
ISBN 978-3-032-16497-1 ISBN 978-3-032-16498-8 (eBook)
https://doi.org/10.1007/978-3-032-16498-8

This Springer imprint is published by the registered company Springer Nature Switzerland AG
The registered company address is: Gewerbestrasse 11, 6330 Cham, Switzerland

To my beloved parents, who have always nurtured my intellectual growth, inspired my creativity, encouraged me to pursue my dreams, and instilled in me the values of hard work and perseverance. To my father, for continually sharing insightful perspectives on the latest trends in marketing and sustainable development. To my wife and daughter, for their understanding, patience, and support throughout my academic career.

Acknowledgement In the preparation of the case studies, as well as some of the discussion questions and illustrative examples in this textbook, I made use of generative artificial intelligence tools, including the ChatGPT, Gemini and Copilot. These tools were employed as part of an exploratory and iterative writing process—to assist with idea generation, language refinement, and the synthesis of publicly available information. All outputs from these tools were critically reviewed, edited, and contextualized by the author to ensure accuracy, originality, and alignment with the intended educational objectives.

Competing Interests The author has no competing interests to declare that are relevant to the content of this manuscript.

Contents

About the Author

Dobromir Kirilov Stoyanov holds a Ph.D. in marketing from the University of Economics—Varna, Bulgaria. He is an Associate Professor at the EM-Strasbourg Business School, University of Strasbourg, France, lecturing a wide range of marketing disciplines, including Retail marketing and logistics, Sustainable marketing, B2B marketing, Advanced international marketing, and Sustainable Customer Relationship Management. Specifically, his research interests are in the fields of vending retail marketing, sustainable marketing and chemistry of marketing. He is a member of the HuManiS research center (UR 7308). He has published papers in the *International Journal of Retail and Distribution Management, Journal of Consumer Affairs and Journal of Macromarketing*. He is also the author of the textbook *Vending Retail Marketing: Text and International Cases* (2025), published by Springer Nature.

The Call for Sustainable Future: A Fad or a Must?

1

Source Stock image from MS Word

Chapter Introduction

Despite the multitude of publications emphasizing the importance of embracing the sustainable marketing philosophy by businesses, governments, and individuals, there still exist skeptics who resist recognizing the urgent need for a shift in human behavior. Moreover, in the digital age where fake news can emerge from nowhere and diffuse everywhere at the speed of light, sometimes it is not so obvious for the general public how to distinguish a conspiracy theory from the reality. This challenge is particularly pronounced in countries where trust in conventional media (such as magazines, radio, and television) is declining, and national legislative bodies are not taking robust measures to control the dissemination of unverified information. Therefore, it is crucial to counter skeptics' arguments and provide readers with insightful tools for making informed decisions about what to believe. This chapter

D. K. Stoyanov, *Sustainable Marketing*, Springer Texts in Business and Economics,
https://doi.org/10.1007/978-3-032-16498-8_1

seeks to elucidate why marketers, now more than ever, must adopt the genuine values of the sustainable development concept and how consumers can improve their ability to discern credible information. In doing so, it aims to empower individuals to navigate the complexities of the digital landscape and contribute to a more informed and sustainable society.

1.1 Understanding Sustainability Skepticism: Approaches to Revealing the Truth

According to renowned astrophysicist Neil deGrasse Tyson (2024), there are three distinct categories of truth that influence how we perceive and interpret reality:

1) *Personal truth:* it refers to beliefs and perspectives shaped by an individual's experiences, emotions, and perceptions. It is subjective and varies from person to person. For example, someone might believe, "Winters are still cold where I live, so climate change isn't real." This reflects confirmation bias—relying on personal experience to dismiss broader scientific trends. However, climate change refers to long-term shifts in global climate patterns, not short-term weather fluctuations. As Elon Musk used to say: "one of the biggest mistakes people generally make, and I'm guilty of it too, is wishful thinking. You know, like you want something to be true, even if it isn't true. You ignore the real truth, because of what you want to be true. This is a very difficult trap to avoid" (Wilson, 2020).
2) *Political truth:* political truth consists of ideas that gain acceptance through social consensus, power structures, or ideological influence. These truths may not necessarily be factually accurate but become widely accepted due to repetition, authority, or group belief. For instance, a politician or corporation might claim, "Our country is a leader in sustainability because we planted trees," while still promoting deforestation, fossil fuel use, or weak environmental policies. This is an example of greenwashing, where selective truth is used for political gain rather than meaningful environmental action. A government may also argue, "We can't afford to transition to renewable energy because it will hurt jobs," ignoring the fact that the renewable sector is rapidly creating new employment opportunities. This reflects status quo bias, where short-term economic concerns are prioritized over long-term sustainability.
3) *Objective truth:* it exists independently of personal beliefs or political influence. It remains valid regardless of whether people accept it. For example, climate change is real and driven by human activity due to greenhouse gas emissions. Regardless of personal beliefs or political agendas, overwhelming scientific evidence confirms that global temperatures are rising, extreme weather events are becoming more frequent, and ecosystems are being disrupted. Despite attempts to manipulate public opinion, recent studies published in high-impact-factor journals have debunked the fabricated notion that climate change is a natural

or uncertain phenomenon (Anderegg et al., 2010; Carlton et al., 2015; Cook et al., 2013). By analyzing large samples of peer-reviewed academic papers, these studies provide substantial evidence that climate change is indeed an anthropogenic phenomenon. The sporadic appearance of conflicting findings is often attributable to issues of validity, reliability, or methodological limitations rather than genuine scientific uncertainty. For example, Diethelm and McKee (2009) classify the main science denial practices into the following interrelated categories:

- *Fake experts*: this category includes individuals who pretend to be experts on a particular topic such as medicine, health nutrition, or sustainable development, but share extremely unconventional views to the established knowledge in the field. For example, in the 1980s Phillip Morris along with the US Tobacco Institute and a group of lawyers launched the European Consultancy Programme. This project, initially established as secret one, involved hiring scientists to find proofs that second-hand smoke is harmless and therefore to restore smoker confidence and defend the interests of the tobacco industry. A similar approach was followed by the American Petroleum institute that in the late 1990s recruited pseudo-experts to convince the industrial stakeholders that global warming does not represent a serious threat to the humanity and the other ecosystems, and therefore does not require immediate measures.
- *Logical fallacies:* these are misrepresented events used for commercial and political interests, initially appearing plausible but lacking robust scientific support. Logical fallacies in this context encompass *red herrings*, which involve intentional efforts to divert the argument, and *straw men*, wherein the opposing argument is misrepresented to facilitate easier refutation. Logical fallacies share similarities with quoting someone out of context but on a larger scale. In quote mining, the original context of a message is removed to mislead the audience about its intended meaning. The main concern here does not arise from simply removing a quote from its original context, as quotations inherently encompass this, but from the sender intentionally opting to exclude particular information that could enhance the receiver's accurate understanding of the actual intentions behind the chosen words. For instance, certain pro-smoking groups employed anti-smoking claims by highlighting Hitler's support for the anti-smoking movement in Nazi Germany. In 1949, the cigarette brand Camel even launched an advertising campaign ("What cigarette do you smoke, Doctor?") revealing study results that claimed "more doctors smoke Camel than any other cigarette," intentionally implying that cigarettes are not as harmful as commonly believed. Similarly, climate change skeptics often argue that if global temperatures were rising universally, winters should not be so cold in certain geographical regions.
- *Impossible expectations:* this denial approach involves setting unreasonable expectations to the public of what a scientific study can deliver as results. In an environmental context, a notable instance of applying this strategy is evident in

claims that climate change is a natural, not human-induced, phenomenon. This argument suggests that inadequate temperature records for older time periods make it impossible for researchers to develop precise models for accurately measuring humanity's actual impact. In a social context, the tobacco industry used the manipulative practice of unrealistic expectations to make the previous century consumers believe that smoking does not cause cancer and thus public authorities should not take strict and immediate measures against it. In 1969, Brown & Williamson Tobacco corporation introduced the following advertising campaign: "The cigarette industry is being maliciously, systematically lynched. Who is to say it won't happen elsewhere?......Ten years ago, there was a cancer scare over the wax in milk cartons. And overusing iodine to get a suntan. These theories were about as valid as the one that says toads cause warts. And they're about as valid as today's scare-tactics surrounding cigarettes. Because no one has been able to produce conclusive proof that cigarette smoking causes cancer. Scientific, biological, clinical, or any other kind. It's more than cigarettes being challenged here. It's freedom. We will continue to bring to the American people the story of the cigarette and any other legal product based upon truth and taste. We believe that free speech and fair play are both the heritage and promise in our society of free and responsible enterprise" (Glantz et al., 1996, pp. 185–86).

- *Selectivity* or *Cherry-Picking:* this misleading practice relies on the intentional selection of specific information, data, or examples that support a particular viewpoint while disregarding other relevant information that might contradict or weaken that viewpoint. It involves a biased representation of reality with the aim of manipulating or influencing perceptions. This strategy is rooted in the use of partial scientific evidence—choosing atypical or outdated studies or models to challenge the prevailing consensus or highlight flaws in the least robust papers supporting it, all with the goal of undermining the entire field. The process includes oversimplifying the understanding of well-known phenomena or jumping quickly to conclusions based on a limited amount of evidence, which can sometimes lead to absurd results. A well-known myth, grounded in another cognitive bias, asserts that if something cannot be measured, it cannot be managed. But is this assertion truly accurate? We hold the belief that, from a statistical perspective, almost anything can be proven and may appear scientifically correct, even though the actual reasons might be that the research model applied in the study is not sufficiently valid or reliable. As the famous phrase—popularized by Mark Twain in his Autobiography—states, "There are three kinds of lies: lies, damned lies, and statistics." (University of York, n.d.). For instance, the absence of rigorous methodology can be attributed to a *spurious correlation*, a type of correlation that emerges when two variables have no discernible causal relationship in real life but can be statistically linked due to each being separately related to another variable known as a "confounding variable" (See Fig. 1.1). For instance, analyzing data for a long enough time period, we can falsely assume that as there is a strong negative relationship one might erroneously conclude that an apparent strong negative relationship between global average temperatures and the number of pirates traversing the oceans implies that increasing piracy in the twenty-first century would be an effective strategy to combat global warming while in fact each of them are most likely related to the industrialization era.

Hence, it is crucial to bear in mind that correlation does not imply causation, and apparently unrelated factors can exhibit statistical relationships purely by chance—a concept largely popularized by Vigen (2015, 2024) through his book and website. As an illustration, he pointed out correlation coefficients, such as the one between US spending on science, space, and technology and the number of suicides committed by hanging, strangulation, and suffocation ($r = 0.99208$), and another between per capita consumption of Mozzarella cheese and the number of civil engineering doctorates awarded ($r = 0.95864$), using reliable data sources. Moreover, with the rise of Big Data and cloud computing technologies, scientists now have access to global databases, enabling them to manipulate vast amounts of data and swiftly identify statistically significant patterns that may not necessarily manifest in the real world. Calude and Longo (2017) raised the awareness about the potential threats to the value of information derived from randomly generated and sufficiently large databases, suggesting that the majority of correlations within such datasets are likely spurious. This explains how and why even complex mathematical models, often perceived as a benchmark for academic quality, might also be nowadays biased (i.e., whether intentionally or not), which means that people more than ever before need strong soft skills (e.g., analytical and critical thinking) to see beyond the visible tips of the statistical/methodological iceberg.

Another factor contributing to the conflicting results in studies on sustainable development could be attributed to the phenomenon known as "*Simpson's paradox*." This statistical anomaly occurs when a trend or correlation observed within individual groups of data disappears or even reverses when these groups are combined. Often, this paradox arises when frequency data are mistakenly interpreted as causal relationships. Resolving Simpson's paradox requires careful consideration of confounding variables and causal relations in statistical modeling. By appropriately addressing these factors, researchers can untangle complex data sets and arrive at more accurate conclusions. An example of how Simpson's paradox could be misinterpreted in the context of sustainable development denialism might involve analyzing carbon emissions data from different industries. Let's consider two industries: heavy manufacturing and renewable energy production. Individually, it may seem that heavy manufacturing industries have made significant strides in reducing their carbon emissions over time, while renewable energy production has shown a less impressive decline. However, when combining the data from both industries, it might appear that overall carbon emissions have decreased only marginally, or perhaps even increased. A sustainable development denialist might misuse this observation to argue that efforts to transition to renewable energy sources are ineffective or unnecessary. They could claim that despite the apparent improvements in heavy manufacturing, the continued reliance on renewable energy

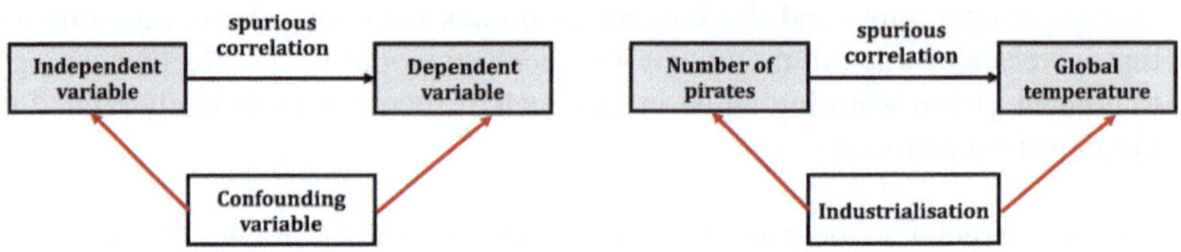

- Independent variable: variable being manipulated to measure its impact on a dependent variable (Factor)
- Dependent variable: a variable that may change in response to changes in other variables (Outcome) .
- Confounding variable: hidden factors for the researcher that might have an impact on the relationship (positive or negative).

Fig. 1.1 Spurious correlation framework

hasn't led to a substantial reduction in overall emissions. However, the paradoxical nature of the data could be explained by the disproportionate contribution of heavy manufacturing to total emissions compared to renewable energy production. While heavy manufacturing industries have made progress in reducing emissions on a per-industry basis, their overall emissions might still outweigh the relatively smaller emissions from renewable energy production. Thus, misinterpreting Simpson's paradox in this context could lead to misguided conclusions and undermine efforts to combat climate change and promote sustainable development.

4) *Conspiracy theories* often challenge the credibility of well-established academic journals by contending that their entire peer-reviewed publication process is biased. These theories argue that the process favors scientific publications supporting specific evidence while rejecting impartial opinions from independent researchers who may have arrived at opposing conclusions. For instance, the tobacco companies accused the anti-smoking movement in secret agreements with other stakeholders that aim to undermine the industry by delivering false evidence about the negative effects of smoking on human health. This approach resembles a lot to a situation in which a thief screams "catch the thief."[1] Some proponents of conspiracy theories may assert claims such as that

[1] This is a Chinese proverb implying that a person who accuses others of wrongdoing may be the actual wrongdoer. It is applied in situations where someone seeks to conceal their illicit actions by shifting blame onto others. The origin of this proverb can be traced back to an ancient fable recounting the tale of two villages at war separated by a river. On the west side of the river there was a trader who was selling simultaneously military equipment to both sides—swords to his own villagers and shields to villagers living on the east side of the river. Despite the profitability of his business for many years, when the neighboring villages decided to make peace, the trader, who found this unfavorable, falsely accused the eastern village of attacking his own, reigniting the conflict. The same seller had a friend who was living on the opposite side of the river and was the only one thief in the entire region, as a result of which no one could catch him as according to his words it took a thief to catch a thief. However, one day the seller understood actually why his friend could never be caught by the authorities—because every time he was stealing from the others, he was pointing out someone else and therefore diverting the attention from him. This ancient

"global warming is a multi-billion-dollar worldwide industry, created by fanatically antiindustrial environmentalists" (as cited in Leka & Furnham, 2024, p. 4) or that it was "created by and for the Chinese in order to make U.S. manufacturing noncompetitive" (Trump, 2012). For instance, Godin (2011) introduced the concept of the "marketing of conspiracy theories," emphasizing the importance of recognizing that "facts don't captivate us, theories do… What is being marketed is doubt, not evidence. Doubt is a commodity that people often desire, especially when it provides them with a sense of comfort."

1.2 Experiencing Sustainability: An NLP Perspective

Another possible explanation involves understanding how our brain is programmed to process verbal and nonverbal information from the external world. According to that model, through our five senses we continuously receive a large amount of information (million and even billions of bits per second). Relying on all our five senses (i.e., visual, auditory, olfactory, kinesthetic, and gustatory) to make sure that what we believe to be true in life is what we actually experience through our sensory receptors. So, let's take a look at some examples of sensual experiences that prove to us that even without scientific evidences if we carefully observe ourselves, the people, and environment around us, we can still have a pretty good idea of the state at which our society and ecosystems are (See Fig. 1.2)

Of course, the sensory observations presented in Fig. 1.2 are just a few simplistic examples showing that even without having access to scientific studies, human beings are able to feel the social and environmental changes that have been happening around them only if they start paying attention to the totality of the information that they receive from the world. Some sustainability skeptics might say that things are never as easy as they might look, and they might be partially right. Winston Churchill once said that "democracy is the worst form of Government except all those other forms that have been tried from time to time" (UK Parliament, 1947). The same thing we can say about our five senses. They have their limitations and can easily mislead us (e.g., just google some of the following terms: optical, auditory, olfactory, tactile, or taste illusions to understand how unexpectedly fragile to manipulations are the five senses). For instance, think about some of the famous magician tricks performed in Las Vegas by David Copperfield or Penn and Teller—"the closer you look, the less you see." Probably in the future more studies will prove that as human beings, we have more than five senses (e.g., the so-called "extrasensory perceptions") that can help us make more reliable estimation of the surrounding environment but for now our main senses stay the key receptors of the incoming information. Therefore, we can never be sure at 100% if what we experience through our five senses is actually real or misleading but

story resonates in the twenty-first century, paralleling instances where political leaders, in collaboration with large corporations, behave disrespectfully toward society and the environment while portraying themselves as victims.

the more consistent is the information that we collect from the different senses the higher will be the probability that what we experience is an actual representation of the reality.

According to the NLP model developed by James and Woodsmall (1998), and presented here from a sustainability perspective, we as humans apply three main filters to incoming information to prevent cognitive overload and to distinguish what is relevant from what is not (see Fig. 1.3).

- *Generalization*—with the help of universal quantifiers (e.g., all, always, never, everyone, no one) we are often inclined to accept that what we experience once or several times is an absolute truth in life. Some examples of skeptical generalizations invoke claims that all NGOs and green activists exaggerate, every company is only interested in generating profits and its social and environmental contributions are nothing but greenwashing, all politicians are corrupted, climate is always changing, humans can never change their bad habits or reverse global warming. Other reality generalizations involve statements that all your senses tell you that climate change is happening, that everyone is causing it and should be held responsible.
- *Deletion*—we concentrate on what is more important at a specific moment while not allowing the rest of the available information to penetrate into our consciousness. Some examples of reality rejection include the consistent denial of climate change as a phenomenon caused by humans without objective consideration or having reliable counter arguments.
- *Distortion*—it occurs when we do not only refuse to accept the totality of the incoming information, but we are also willing to misinterpret it as a result of which we can either over- or underestimate its importance. Some examples of reality misrepresentation involve being too optimistic about the future

Sight (*Eyes*)	**Audio** (Ears)	**Scent** (*Nose)*	**Taste** (Mouth)	**Touch** (*Skin*)
Unusual natural phenomena, including Sahara snow, melting glaciers, rising sea levels, disrupted animal migration, and shifted seasons, along with more frequent and intense disasters like hurricanes, earthquakes, droughts, and floods.	Much of the information we receive daily through mainstream media—TV, radio, and the internet—focuses on social and environmental issues that require collective action to address.	Respiratory problems such as suffocation, coughing, and sore throat are common due to polluted air, dust, and smoke, especially in large cities. Additionally, climate change and the use of hybrid or genetically modified seeds have reduced the fragrance of flowers.	Tasteless hybrid fruits and vegetables cultivated soilessly with significantly reduced content of nutrients and vitamins that can last for weeks without being stored in a fridge.	Increased allergies and skin irritations result from industrial pesticides and chemicals, while pollution causes dry, prematurely aging skin. Acid rain can lead to hair loss, temperature fluctuations contribute to heart attacks, and excessive sun exposure causes sunburns and heatstroke.

Fig. 1.2 A five-senses perspective on the need for a sustainable future

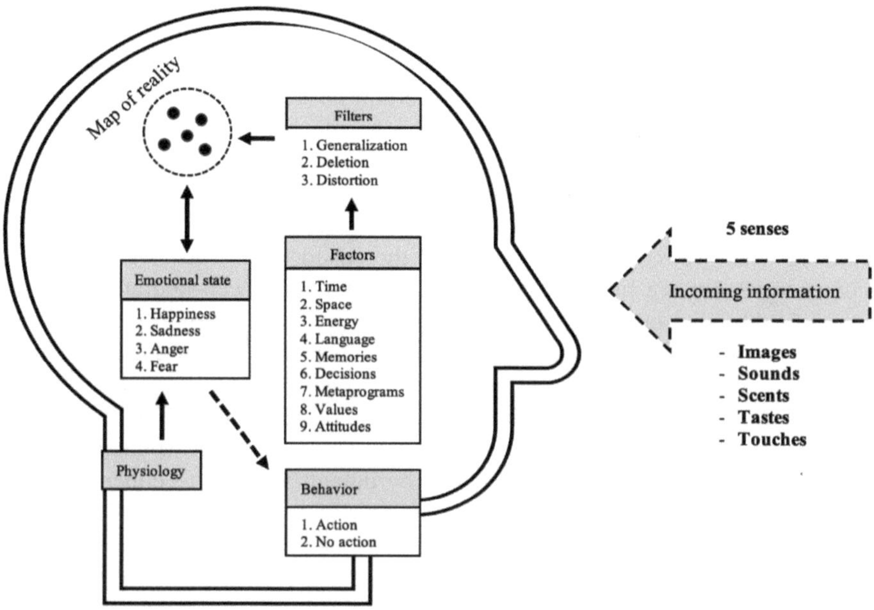

Fig. 1.3 A typical NLP communication model. *Source* Own development based on James and Woodsmall (1998).

regardless of the available clues on the evident deterioration of the ecosystems, compost and recycling procrastination behavior, the attempts of the industrial and political propaganda to undermine public's approval of the need of adopting regulations on the CO_2 emissions, pseudoscience looking for scientific proofs where they do not exist (e.g., health benefits of smoking).

Any piece of information that successfully passes through our filters results in the construction of a thought. Several thoughts on the same topic, when combined, help us form our so-called "internal representations" or "maps of reality," which lead to certain emotional states (i.e., happiness, sadness, anger, fear), which in turn trigger our decisions to behave or not in a particular manner. There are nine main factors that help us determine which filters to apply in any given situation. These are the following ones:

- *Time*—this factor is related to the exact temporal moment when we receive information, the frequency and duration of exposure at which our brain registers the external stimuli. At what time of the day do we receive information? Early in the morning or late in the evening? How often and for how long are our senses bombarded with audio-visual messages? At what moment of our life do we receive the information? When we are young? When we are old? When we are unemployed? When we work as a CEO, president, or green activist? What are our priorities at that particular time period?

- *Space*—this factor is related to the physical place or the environment surrounding us when our senses are exposed to certain informational stimuli. Are we in a noisy or quiet environment? Are we at a hostile new place or at home in our comfortable massaging bed? Is it rainy or sunny? Do we live in a region with high or low risk of natural disasters?
- *Energy*—this factor is related to the physical and mental health state at which we are when we receive a message from the outside world. Are we tired or full of energy? Are we hungry or full? Are we thirsty or well-hydrated? Are we on a healthy diet or are we eating junk food? Are we suffering from a disease or are we in relatively good health? Do we sleep well or not enough? Do we exercise or do we have a sedentary lifestyle?
- *Language*—as human beings when a new piece of information is delivered to us in foreign languages, through slang, complex terminology, or wooden language (sometimes used in political speeches), by default we tend to either ignore or misunderstand the message without being able to properly analyze the value of the information which may lead to confusions and false conclusions.
- *Memories*—this factor is strongly interrelated with two other determinants (i.e., decisions and values and beliefs). It refers to the past individual and collective experiences that have an impact on the way we perceive the reality at the present moment. These memories can be both conscious and unconscious, based on decisions that we took on our own, or experiences acquired through observational learning. How were we educated? What personal example did our parents or teachers give us when we were children? Composting, recycling, and respecting nature? Smoking, throwing cigarette butts and any other non-degradable waste on the ground, laughing at people with disabilities? Did we or some of our relatives suffer in the past from a severe disease?
- *Decisions*—they represent the past decisions about who we are and what we are capable of. They affect the decisions that we are faced with now. What promises did a CEO make to the shareholders? How much money did a political leader raise for his presidential campaign? Which companies or industries are the largest financial contributors? What explicit or implicit agreements did a political leader establish with campaign fundraisers? What promises did a president or prime minister give to the electors? How strong is the level of obligations from previous decisions?
- *Meta programs*—they involve the most unconscious set of psychological characteristics that we were born with and that distinguish us from other individuals and members of the society. They are related to our personality profile (e.g., introvert or extrovert, thinking or feeling, sensing or intuitive; judging or perceptive; optimist, pessimist, realist, or idealist). How do we take our decisions? Mostly based on our intuition, brain, or heart? Do we see the world through a logical lens? Do we look for cause and effect or do we attribute life events to randomness?

- *Values and beliefs*—a set of criteria that determine our perceptions for good and bad, right or wrong (e.g., altruism, responsibility, integrity, diligence, trustworthiness, discipline, justice). They are based on our cultural background, superstitions, or religious beliefs and are formed as a result of the memories, decisions, and metaprograms. Is it acceptable to lie or not? Is it a good thing to help others? Which values should we defend—corporate or human values? Can money buy us everything?
- *Attitudes*—these are values and beliefs that we have about specific things such as companies, brands, and political parties. They can be positive, negative, or neutral and depending on their direction and magnitude they can influence to a different extent our filters to accept, reject, or mispresent the incoming information.

Though we tried to simplify how human brain processes information registered by the five senses, it is important to mention that brain and human receptors also have their own limitations due to which our reality perceptions can be misled. Therefore, it is not always that easy to decide if we should accept, reject, or mispresent any piece of information that reaches our brain. A good reference on this topic might be the science documentary series of National geographic "Brain games" which presents through a variety of scenarios some misconceptions about our brain. As a sequential model, the NLP framework assumes that human behavior can be predicted in the following sequence: thoughts lead to emotions which trigger behaviors, and though in many cases human behavior is not as straightforward as the model suggests, the reality that we all experience is largely determined by what we do inside our heads. This means that everyone experiences any given situation differently. It's our internal representations (subjective perceptions) that determine how we view the world and everything we experience (i.e., if climate change is a hoax or an urgent problem to solve). Nowadays, some people—including political leaders, online bloggers, lobbyists, and even members of the scientific community—refuse to believe in objective facts or deliberately misrepresent them, creating their own perceptual maps of reality because it is more convenient for them. This phenomenon can be described by the term "truthiness," which refers to "the quality of seeming or being felt to be true, even if not necessarily true" (Oxford English Dictionary, 2023). Another term closely related to the previous one and frequently used since Brexit, is "post-truth." This neologism was recognized in 2016 by Oxford Dictionaries as the international word of the year and an essential characteristic of the twenty-first century, as it refers to situations in which personal beliefs or emotions take precedence over objective facts (Oxford University Press, 2016).

1.3 Spotting Fake News on Sustainability at the Digital Age

In the digital age, the ability to critically assess online information, known as digital literacy, is as crucial as traditional literacy skills. As the saying goes, "A lie can travel halfway around the world while the truth is putting on its shoes," a quote often attributed to Mark Twain without substantiated evidence. This aphorism underscores the virality of misinformation in the digital age, where falsehoods can spread rapidly and widely, outpacing efforts to debunk or counteract them. In this era of information overload and algorithmic echo chambers, discerning truth from fiction becomes increasingly challenging, necessitating heightened vigilance and critical thinking skills to navigate the murky waters of online discourse. Digital literacy is particularly pertinent in the context of sustainability. Ensuring the credibility of online sources requires meticulous examination of authorship and sponsorship. Questions arise: Can information be verified through reputable platforms like CNN, BBC, or The New York Times? Are study results readily accessible for methodological scrutiny? Who funded the research, and are there potential conflicts of interest?

Cook et al. (2018) have developed a valuable systematic framework for identifying misleading statements used by climate change denialists, which also has broader applications in sustainability. As presented in Fig. 1.4, the framework consists of the following steps:

1. **Identifying the claim**: begin by recognizing the specific assertion made by the denialist. It is crucial to capture the exact wording of the claim to ensure an accurate evaluation and avoid misinterpretation. For example, if someone argues, *"Renewable energy sources are too unreliable to meet our energy needs,"* this statement serves as the claim requiring examination.
2. **Constructing the argument**: break down the claim into its underlying premises and conclusion. The goal is to outline the reasoning that supports the denialist's statement. For example, if the claim is that *"Renewable energy sources are too unreliable,"* the premises might include points like "*Solar energy only works during the day*" and "*Wind energy depends on weather conditions.*"
3. **Determining inference**: assess the truths of each premise. This involves checking facts or relying on scientific data to confirm the accuracy of each statement. Using the previous example, one might examine the actual data on the efficiency of solar panels during cloudy days or the advancements in energy storage technologies that mitigate the variability of renewable sources.
4. **Checking for logical validity**: analyze the reasoning to ensure that it logically connects the premises to the conclusion. It's crucial to determine if the argument follows a valid structure. For instance, if the premises are true but do not logically lead to the conclusion that the renewable energy sources cannot fulfill energy demands, then there is a flaw in reasoning.

   4a. Identifying hidden assumptions: look for any unstated assumptions that the argument relies on. These assumptions may not be explicitly stated but can

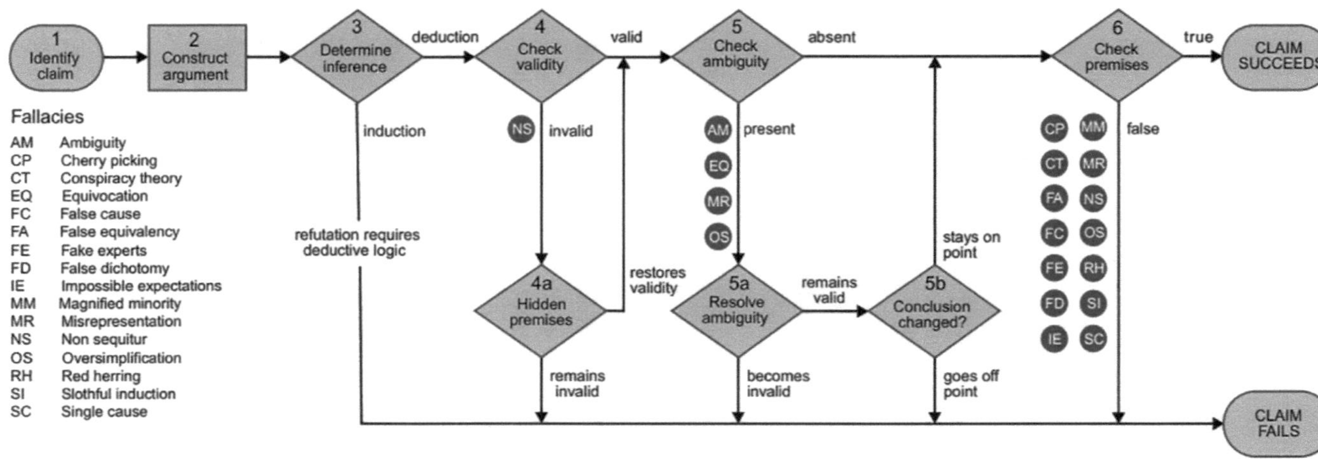

Fig. 1.4 A framework for evaluating the quality of information about climate change. *Source* (Cook et al., 2018, p. 3). Licensed under CC BY 3.0.

significantly impact the argument's validity. For example, the assumption might be that all sources of energy must operate at full capacity at all times, ignoring the role of a mixed energy portfolio.

5. **Checking for ambiguity**: investigate any vague language or ambiguous terms used in the argument. Ambiguities can mislead or confuse the interpretation of claims. If the assertion includes terms like "unreliable," one should clarify what "unreliable" means in the context of energy generation—whether it refers to capacity, consistency, or affordability.
6. **Check premises for truth or plausibility**: Once you've confirmed that an argument is logically valid—meaning the conclusion follows from the premises without ambiguity—the next step is to evaluate whether the premises themselves are actually true or plausible. This is essential because even a logically sound argument can lead to a false conclusion if it's built on shaky or incorrect foundations. Importantly, many denialist arguments collapse not due to complex science, but because they rely on clearly false or highly questionable assumptions. For instance: The claim "If something caused an effect in the past, it will always be the cause in the future" is implausible when multiple causes are possible. Similarly, "If there's no empirical evidence for something, then it isn't happening" confuses the absence of evidence with evidence of absence—a basic logical error.

The model discusses various fallacies that exemplify errors in reasoning commonly found in climate change arguments, contributing to misunderstanding and misinformation on the topic.

- **Ambiguity (AM)** arises when a word or phrase is used in a way that makes its meaning unclear or open to multiple interpretations. An example of this fallacy can be seen in the statement, "climate change is harmful." Without further clarification, it is difficult to determine whether the speaker is referring to the impacts on human health, ecosystems, or economic systems, leading to confusion about the argument's intent and validity.
- **Cherry picking (CP)** involves selectively presenting evidence that supports a claim while ignoring relevant data that might contradict it. For instance, a climate change skeptic may highlight a single year with unusually cold temperatures as evidence that global warming isn't occurring. This cherry-picked information overlooks the broader context of decades of data showing a clear upward trend in average global temperatures.
- **Conspiracy theory (CT)** is a fallacy that suggests a secret group is manipulating events or information for nefarious purposes, often without substantial evidence. An example would be claims that climate scientists are part of a conspiracy, receiving government funding to promote climate change to profit from grants and research opportunities. These assertions typically lack credible evidence and distract from the scientific consensus on climate change.

- **Equivocation (EQ)** occurs when a term is used in two different senses within an argument, leading to misleading conclusions. A classic example is the argument: "Nothing is better than eternal happiness; a ham sandwich is better than nothing. Therefore, a ham sandwich is better than eternal happiness." In this case, the word "nothing" is used ambiguously, causing the conclusion to lose its validity.
- **False cause (FC)** assumes a cause-and-effect relationship between two events without sufficient evidence. For example, one might argue that since climate change began, there has been an increase in allergy diagnoses, suggesting that global warming is directly causing more allergies. This reasoning neglects other possible factors that could affect allergy rates, such as increased pollution or changes in lifestyle.
- **False equivalency (FE)** asserts that two arguments or situations are equivalent when they are not, ignoring significant differences. For example, treating climate change activism and climate change denial as equally valid viewpoints overlooks the overwhelming scientific consensus and evidence supporting anthropogenic climate change. This fallacy fails to consider the weight of evidence on one side versus the other.
- **Fake experts (FE)** refer to citing individuals as authorities who either lack relevant expertise or have been discredited in their fields. An instance of this would be referencing a poorly qualified scientist who disputes established climate science based on limited or outdated knowledge. By promoting such fake experts, the argument undermines the validity of scientific discourse.
- **False dichotomy (FD)** presents a situation as having only two alternatives, effectively ignoring other possible options. For example, one might argue that society must either completely abandon fossil fuels or accept catastrophic climate change. This simplifies the complexity of environmental policy by ignoring potential compromises and varied strategies for reducing emissions.
- **Impossible expectations (IE)** are fallacies that place unrealistic demands on evidence in order to dismiss valid arguments. An example could be stating that climate scientists cannot definitively prove that human activity is the only cause of climate change, thereby negating the strong evidence supporting human influence. Such a stance creates a barrier to understanding the multifactorial nature of climate dynamics.
- **Magnified minority (MM)** gives undue emphasis to a small minority of dissenting opinions while ignoring the majority consensus. A common occurrence is when media outlets spotlight one or two scientists who question climate change, while downplaying the thousands of researchers who support it. This misrepresentation creates a misleading perception of the debate's balance.
- **Misrepresentation (MR)** distorts an argument or data to make it easier to attack or refute. For instance, if someone claims that climate scientists say all parts of the world will experience the same degree of warming, they misrepresent the nuanced discussions and findings, which often recognize regional variation and complexity.

- **Non sequitur (NS)** involves conclusions that do not logically follow from the premises provided. For instance, asserting that "if we stop using fossil fuels, it will lead to economic collapse" presents a conclusion that doesn't logically follow from the premise about reducing fossil fuel usage. This disconnect calls into question the validity of the argument made.
- **Oversimplification (OS)** reduces complex issues to overly simple terms, ignoring important information and nuances. For example, stating, "Climate change is just about reducing CO2 emissions" overlooks the broader implications, including impacts on ecosystems, social dynamics, and the economy. This simplification can hinder meaningful discussion about necessary actions for mitigation.
- **Red herring (RH)** introduces irrelevant information to divert attention from the main argument or issue being discussed. An example could be in a debate about climate change impacts, where one party introduces historical weather events that are unrelated to current climate trends, effectively steering the discussion away from pertinent topics.
- **Slothful induction (SI)** occurs when someone ignores significant evidence in favor of a counterclaim that has little to support it. For example, dismissing overwhelming scientific studies linking emissions to climate change while only citing one isolated cold year is a form of slothful induction. This fallacy shows a failure to adequately engage with the broader body of evidence.
- **Single cause (SC)** refers to attributing an effect to a single cause when multiple factors may contribute to it. An illustration of this fallacy is claiming that increased hurricane activity is solely due to climate change without considering other possible contributing elements such as natural climatic variability. This oversimplifies the complexity of meteorological phenomena.

Chapter Conclusion

To address the earlier question posed in this chapter about whether a sustainable future is a passing trend or an imperative, we can conclude that not only should companies be concerned, but the entire society—comprising individuals, businesses, and governmental authorities—must embrace change and foster greater responsibility. Global warming is not a mere urban legend and climate change is not a game with a reset button; it functions more like a ticking time bomb, constantly sending us warning messages. Some individuals have already recognized this and taken actions to preserve resources. However, the urgent need for change should not come at the cost of spreading social disinformation. Despite the existence of various space projects by NASA and SpaceX, earth remains the only habitable planet within our reach. While there may be aspirations to colonize other planets in the galaxy, we should remain vigilant about our daily behaviors. The earth is not expendable; it is crucial to ensure that our children inherit a healthy planet.

Discussion Questions and Practical Assignments

1. The United Nations Framework Convention on Climate Change (UNFCCC) aims to develop viable and collaborative solutions between countries in the fight against the negative consequences of climate change. Since 1992, UNFCCC member states meet annually to discuss the possible improvement measures that each of them can adopt to combat climate change.

   Make a short table to draw the timeline for at least the last 10 years and identify the main decisions taken during the latest UN climate change conferences (See Table 1.1). *Effectiveness* is often defined as *doing the right things*, and efficiency as *doing things right.* Do you think that UNFCCC has been effective and efficient in the fight against climate change and global warming? Discuss critically the topic providing arguments for both perspectives. Do you think that the temporary withdrawal of the USA from the Paris agreement (COP23) from 2017 to 2021) had some positive and/or negative consequences to the USA and or the world? Why? Why not? What do you think are the next logical steps that countries should undertake in the following climate change conferences to more effectively and efficiently fight against challenges of climate change?
2. Read the paper by Cook et al. (2018), titled "Deconstructing climate misinformation to identify reasoning errors," published in *Environmental Research Letters*, Vol. 13 No. 2, pp. 1–7, which is available on open access online. Seek to comprehend the various types of fallacies outlined and learn how to easily identify them. Subsequently, explore social networks for instances of social or environmental fallacies and attempt to assess the validity of their claims (e.g., HAARP as an alleged ultimate climate change weapon, injecting bleach as a cure for treating COVID-19, or Earth's climate changing naturally without human impact). Consider, whether the model could reliably detect deepfakes. If not, what alternative suggestions would you propose for the new generation of internet users who are just starting to navigate the online world? Share your critical insights on how to enhance digital literacy, specifically in identifying deepfakes.
3. Regardless of the gaining importance of sustainability issues and the demarketing measures of policymakers in developed countries the global sales revenues of the tobacco industry are still rising. Why? Explain this phenomenon?
4. Use the Neuro-linguistic programming (NLP) communication model presented in this chapter to explain why some people still consider climate change as unreal and share false information about it. Take the example of one particular person

Table 1.1 A framework for evaluating the quality of information about climate change

Climate change conference	Date	Venue (city, country)	Main achievements in the fight against climate change
COP1			
COP2			
COP3			
…..			
COPn			

(e.g., a journalist, scientist, influencer, online blogger, political leader, etc.). What is/are the filter(s) and factors influencing that perception and following behavior?
5. Deepfakes emerged in the late 2010s. The term "deepfake" itself is a fusion of "deep learning" and "fake," reflecting the technology's foundation in artificial intelligence and machine learning algorithms, enabling the creation of hyper-realistic videos, images, or audio files, often indistinguishable from genuine content. While initially gaining attention for its potential in entertainment and digital art, deepfakes have raised significant concerns regarding their misuse. From spreading misinformation to fabricating compromising materials, the rise of deepfakes poses serious ethical, legal, and security challenges.

On April 14, 2020, during the initial wave of the COVID-19 pandemic, Extinction Rebellion, an international environmental movement, circulated a deepfake video through its various social media channels. The video purported to show the Belgian prime minister, Sophie Wilmès, linking the emergence of Coronavirus to the ongoing ecological crisis. Before being debunked as a deepfake, the video had already been viewed by over 100,000 internet users. This incident underscores the risk that even trusted NGOs may be willing to compromise their credibility in pursuit of influencing public opinion on sustainability-related issues. This example serves not only as proof that the road to hell may indeed be paved with good intentions but also highlights the need for greater awareness and regulation around deepfakes to prevent their exploitation for misleading or manipulative purposes.

- Do the ends justify the means? Would you support a sustainable brand or an NGO that has used deepfakes in marketing campaigns?
- Could widespread exposure to deepfakes lead to a "post-truth" society where people reject all information as potentially fake? What are the dangers of such skepticism?
- How can consumers become more media literate? What strategies should brands and educators use to help people critically assess deepfake content related to sustainability?
- Should governments implement stricter regulations on deepfake content? If so, what kind of penalties should apply to those who create and spread deceptive deepfakes?
- Should there be an international legal framework for combating deepfake misinformation or should regulations be left to individual countries?
- What role should social media platforms play? Should platforms like YouTube, Instagram, and TikTok be responsible for policing deepfake content related to sustainability, or is that a slippery slope?
- Given the rapid advancement of AI, is it even possible to develop foolproof detection systems for deepfakes? What are the limitations of current detection methods?

Acknowledgment ChatGPT, an AI language model developed by OpenAI, was used in this chapter to support the summarization of the framework for evaluating the quality of information about climate change developed by Cook et al. (2018). The author has critically reviewed, verified, and adapted the outputs to ensure accuracy, relevance, and alignment with the academic objectives of the work.

References

Anderegg, W. R. L., Prall, J. W., Harold, J., & Schneider, S. H. (2010). Expert credibility in climate change. *Proceedings of the National Academy of Sciences of the United States of America, 107*(27), 12107–12109. https://doi.org/10.1073/pnas.1003187107

Calude, C. S., & Longo, G. (2017). The deluge of spurious correlations in big data. *Foundations of Science, 22*(3), 595–612. https://doi.org/10.1007/s10699-016-9489-4

Carlton, J. S., Perry-Hill, R., Huber, M., & Prokopy, L. S. (2015). The climate change consensus extends beyond climate scientists. *Environmental Research Letters, 10*(9), 094025. https://doi.org/10.1088/1748-9326/10/9/094025

Cook, J., Ellerton, P., & Kinkead, D. (2018). Deconstructing climate misinformation to identify reasoning errors. *Environmental Research Letters, 13 (2), 024018,* doi:https://doi.org/10.1088/1748-9326/aaa49f.

Cook, J., Nuccitelli, D., Green, S. A., Richardson, M., Winkler, B., Painting, R., Way, R., Jacobs, P., & Skuce, A. (2013). Quantifying the consensus on anthropogenic global warming in the scientific literature. *Environmental Research Letters.* 8, 024024. http://dx.doi.org/10.1088/1748-9326/8/2/024024

Diethelm, P., & McKee, M. (2009). Denialism: What is it and how should scientists respond? *European Journal of Public Health, 19*(1), 2–4. https://doi.org/10.1093/eurpub/ckn139

Glantz, S., Slade, J., Bero, L., Hanauer, P., & Barnes, D. (1996). *The cigarette papers.* University of California Press.

Godin, S. (2011, November 23). *The marketing of conspiracy theories.* Seth's Blog. https://seths.blog/2011/11/the-marketing-of-conspiracy-theories/

James, T., & Woodsmall, W. (1998). *Time line therapy and the basis of personality.* Meta Publications.

Leka, J., & Furnham, A. (2024). Correlates of climate change skepticism. *Frontiers in Psychology, 15*, 1328307. https://doi.org/10.3389/fpsyg.2024.1328307

Oxford University Press. (2016). *Word of the year 2016: Post-truth.* https://languages.oup.com/word-of-the-year/2016/

Oxford English Dictionary. (2023, July). *Truthiness.* Oxford University Press. https://www.oed.com/dictionary/truthiness_n

Trump, D. J. [@realDonaldTrump]. (2012, November 6). *The concept of global warming was created by and for the Chinese in order to make U.S. manufacturing non-competitive* [Tweet]. X. https://x.com/realDonaldTrump/status/265895292191248385

Tyson, N. deGrasse. (2024). *Neil deGrasse Tyson teaches scientific thinking and communication.* MasterClass. https://www.masterclass.com/classes/neil-degrassetyson-teaches-scientific-thinking-and-communication

UK Parliament. (1947, November 11). House of Commons debates, Volume 444: Commons Chamber [Official transcript]. *Hansard.* https://hansard.parliament.uk/commons/1947-11-11/debates/ac22a8ea-e202-4be6-b2b6-a5a448a12182/CommonsChamber

University of York. (n.d.). Lies, damned lies and statistics. *History of Statistics.* https://www.york.ac.uk/depts/maths/histstat/lies.htm

Vigen, T. (2015). *Spurious correlations.* Hachette Books.

Vigen, T. (2024). *Spurious correlations.* Retrieved from https://www.tylervigen.com/spurious-correlations

Wilson, J. N. (2020, September 8). *The allure of wishful thinking*. Medium. https://medium.com/change-becomes-you/the-allure-of-wishful-thinking-c7be4b6b6831

2 Sustainable Marketing Process: A Value-Based Approach

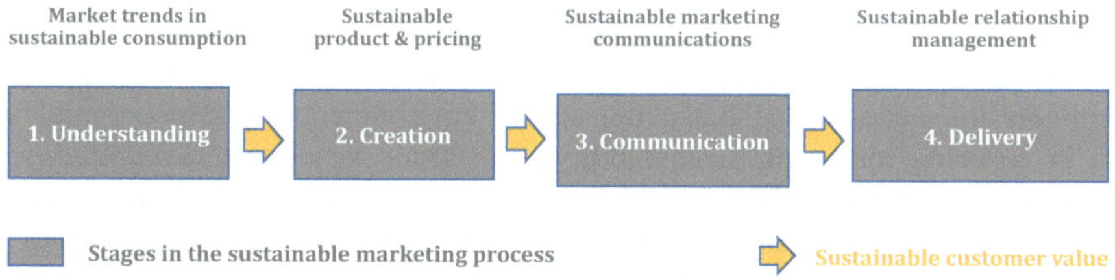

Chapter Introduction

This chapter presents two complementary approaches to sustainability. The first, a structural or systematic approach, adopts a relational perspective and considers how the three main stakeholders in sustainable development—society, businesses, and policymakers—can engage in a value co-creation process. This approach emphasizes the interconnectedness of these stakeholders, highlighting their shared responsibility in promoting sustainability. The second approach, which will be further elaborated in the following chapters, takes a dynamic perspective, viewing sustainable marketing from a linear standpoint as the process of understanding, creating, communicating, and delivering sustainable customer value. This perspective focuses on the sequential steps businesses undertake to embed sustainability into their marketing practices, from identifying consumer needs and preferences for sustainable products to designing and delivering value propositions that align with environmental and social responsibility. Together, these approaches provide a comprehensive framework for sustainable marketing, balancing systemic collaboration with strategic implementation to achieve long-term environmental, social, and economic benefits.

D. K. Stoyanov, *Sustainable Marketing*, Springer Texts in Business and Economics,
https://doi.org/10.1007/978-3-032-16498-8_2

2.1 Sustainable Development: A Stakeholder Approach

One of the earliest and most widely accepted definitions of sustainable development is "development that meets the needs of the present without compromising the ability of future generations to meet their own needs" (United Nations World Commission on Environment and Development 1987, p. 43). Complementing this, the *triple bottom line* approach to sustainable development conceptualize it through three core dimensions: *people* (social equity and well-being), *planet* (environmental stewardship), and *profit* (economic viability) (Elkington, 1997). This approach highlights the necessity of balancing social, environmental, and economic factors in ensuring long-term sustainability. In essence, *economic sustainability* + *environmental sustainability* + *social sustainability* = *sustainable development*. *Economic sustainability* involves the efficient use of resources by businesses or nations to ensure long-term profitability and operational continuity. Without consistent profit generation and responsible resource management, a business cannot sustain its operations; inefficient or irresponsible use of resources undermines its long-term viability (*Profit*). *Environmental sustainability* focuses on reducing environmental impact through energy efficiency, conservation of natural ecosystems, and responsible resource use. This includes addressing issues such as material scarcity, environmental degradation from resource extraction, and embracing circular economy principles (*Planet*). *Social sustainability* refers to the ability of societies or social systems to maintain and enhance the well-being of their members over time. Ensuring long-term social health at the community, organizational, or national level is crucial for sustainable development (*People*).

As presented in Fig. 2.1, the three main domains where stakeholders can meaningfully contribute to sustainable development are:

1. *Sustainable consumption*: this domain involves the demand side of the sustainability equation, where various members of society influence what is consumed

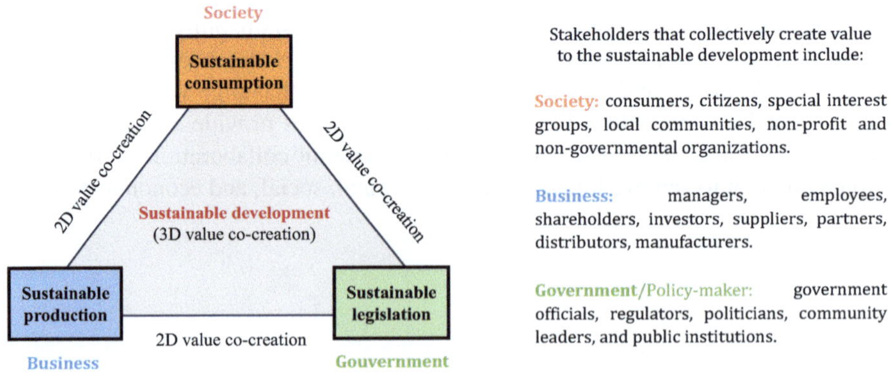

Fig. 2.1 The triangle of sustainable development as a value co-creation process

and how. Stakeholders include consumers, citizens, local communities, NGOs, non-profit organizations, and special interest groups. Their actions—such as making environmentally conscious purchases, advocating for ethical practices, or promoting awareness—shape market demand and societal norms. The topic of sustainable consumption will be explored in depth in Chap. 3: *Understanding sustainable value through consumer behavior*, where we will examine how consumer choices contribute to sustainability, the psychological and cultural drivers behind these choices, and the role of marketing and public policy in facilitating more sustainable consumption patterns.
2. *Sustainable production:* this area pertains to the supply side—how goods and services are created, delivered, and managed across the value chain. It includes businesses and their extended networks: managers, employees, shareholders, investors, suppliers, partners, distributors, and manufacturers. These stakeholders contribute through sustainable sourcing, eco-innovation, waste reduction, circular economy practices, and green supply chain management. For example, Hawken (2010) argues that the primary responsibility for tackling sustainability challenges lies with business corporations, given their influential role in today's world. He emphasizes that every business must address three essential aspects of its operations: *what it takes*—the natural resources and energy it extracts from the environment through activities like mining, logging, and farming; *what it makes*—the goods and services it produces by transforming those resources; and *what it wastes*—the pollution, cultural by-products, and environmental damage it leaves behind. The topic of sustainable production—or, in other words, how businesses can actively contribute to sustainable development—is the primary focus of this textbook. It will be further explored in Chapters 4, 5, 6, and 7, in the context of the processes involved in the creation, communication, and delivery of sustainable customer value.
3. *Sustainable legislation*: this domain involves the formal institutional and regulatory frameworks that enable or constrain sustainability efforts. Key actors include governments, policymakers, regulators, politicians, public organizations, and community leaders. Their contributions include setting environmental standards, creating incentives, enforcing compliance, and shaping public policy in support of long-term sustainability goals. Sustainable legislation refers to laws and regulations that aim to promote the values of sustainable development while balancing the interests of society and the business world. Public policymakers use different levers to achieve these objectives, broadly classified into two categories: *positive measures* and *negative measures* (See Fig. 2.2).

The dominance of either positive or negative measures in national policy often depends on broader cultural factors, such as societal values, the level of trust in government, historical experiences with regulation, the prevailing economic philosophy (e.g., preference for free market solutions versus government intervention), and public attitudes toward sustainability. In cultures with high trust in institutions and strong civic engagement, positive measures may be more effective. In contrast, in societies where compliance is lower or environmental degradation is more

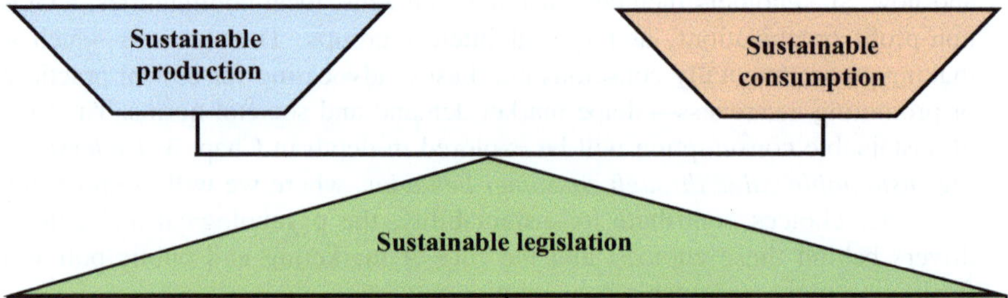

Positive measures	Negative measures
Definition: actions taken by policymakers to encourage responsible behavior among public institutions, businesses, and citizens by offering incentives such as guidelines, subsidies, tax exemptions, and other supportive initiatives aimed at promoting sustainable practices	*Definition:* actions taken by policymakers to enforce responsible behavior through restrictions or penalties such as taxes, fines, or sanctions, targeting public institutions, businesses, and citizens who fail to comply with sustainability standards.
Examples: introducing sustainability guidelines (such as sustainability reporting toolkits to assist firms in disclosing their sustainability practices), establishing voluntary labeling systems, integrating sustainability education into school curricula, and launching public campaigns in collaboration with NGOs to promote sustainable behaviors (e.g., the Green Loyalty Points system in the Netherlands). Governments can also lead by example by adopting sustainable purchasing practices within public institutions, as seen in countries like Belgium, Switzerland, and the UK. Additionally, the concept of tax choice reflects the belief that individual taxpayers should have a say in how their tax contributions are allocated by the state..	*Examples:* sin and luxury taxes, which are designed to reduce the consumption of unhealthy and non-essential products and services, targeting activities with significant negative societal or environmental impacts. These measures include excise taxes on alcohol and tobacco products in the EU, taxes on soft drinks, fast food, and coffee in France, taxes on activities such as gambling, pornography, and prostitution, and taxes on luxury items like yachts, private jets, and jewelry in the USA (1991-1993). These are all Pigouvian taxes, which aim to penalize activities that create externalities—social or environmental costs not typically reflected in market prices.

Fig. 2.2 The balancing role of policymakers: positive and negative levers for sustainability

urgent, negative measures may need to take precedence to achieve rapid results. One of the latest benchmark frameworks related to sustainability is the 17 Sustainable Development Goals (SDGs), developed by the United Nations as part of the 2030 Agenda for sustainable development. These goals aim to address the main global challenges to sustainable development. Many national and regional legislations have been inspired by this framework; for instance, the European Union

Sustainability area	Stakeholders	Commitments	Alignments with the United Nations Sustainable Development Goals (SDGs)
Sustainable consumption	Consumers, NGOs, local communities	Demand for low-emission transport, purchase of electric vehicles, social media posts on sustainable mobility	SDG 3: Good health & well-being SDG 12: Responsible consumption & production SDG 11: Sustainable cities
Sustainable production	Automakers, battery suppliers, tech firms, mobility startups	Design and manufacture of electric vehicles, sustainable batteries and software innovation, green supply chain management	SDG 7: Affordable & clean energy SDG 9: Industry, innovation & infrastructure SDG 8: Decent work & economic growth
Sustainable legislation	Governments, regulators, city councils, public agencies	Subsidies and tax credits for electric vehicles, emission regulations, infrastructure investment	SDG 13: Climate action SDG 16: Peace, justice & strong Institutions SDG 17: Partnerships for the goals

Fig. 2.3 Sustainable value creation in the automotive industry: an SDG's perspective. *Source* The content of this table is partially generated with ChatGPT4.0

requires large companies and listed small and medium-sized enterprises (SMEs) to report detailed information on sustainability matters, aligned with the SDGs. A hypothetical example of the main contributions that society, businesses, and policymakers can make in the automotive industry is presented in Fig. 2.3.

2.2 Stages of the Sustainable Marketing Process: A Value-Based Approach

Sustainable marketing is the process of understanding, creating, communicating, and delivering sustainable customer value (see Fig. 2.4). The following book chapters will explore in detail each of these four stages, but for now, we will primarily focus on the concept of sustainable customer value, which is the cornerstone in the process.

Sustainable customer value can be defined as the subjective ratio or difference between perceived sustainable benefits and sustainable costs (see Fig. 2.5). In simpler terms, before deciding whether to purchase, use, and eventually dispose of a product—whether a physical good or a service—a sustainability-conscious consumer should carefully evaluate all perceived sustainability benefits relative to the associated sustainability costs. These benefits and costs can be conceptualized in multiple ways: (1) from social, environmental, and economic perspectives; (2) across different stages of the consumption process—before, during, and after use

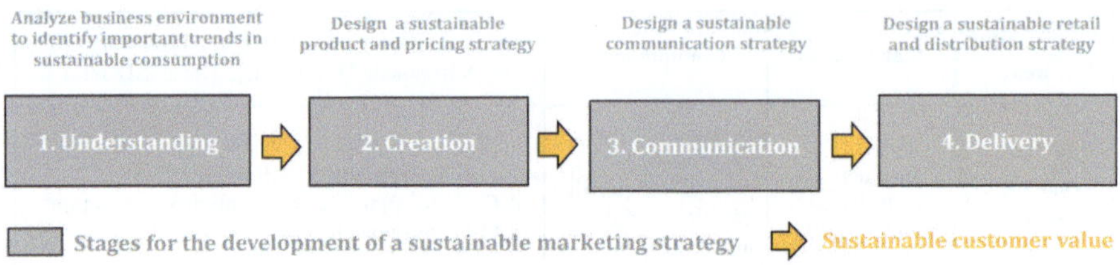

Fig. 2.4 Stages of the sustainable marketing process: a value-based approach

Sustainable customer value

$$\textbf{Sustainable customer value (1)} = \frac{\textit{Sustainable benefits (SB)}}{\textit{Sustainable costs (SC)}}$$

$$\textbf{Sustainable customer value (2)} = \textit{Sustainable benefits (SB)} - \textit{Sustainable costs (SC)}$$

Strategies for adding sustainable customer value

1) Raising sustainable benefits (i. e. SB↑)
2) Reducing sustainable costs (i. e. SC↓)
3) Raising sustainable benefits (SB↑) and reducing sustainable costs (i. e. SB↑ & SC↓)
4) Increasing sustainable benefits more than sustainable costs (e. g. 3SB↑ > 2SC↑)
5) Reducing sustainable benefits less than sustainable costs (e. g. 2SB↓ < 3SC↓)

Fig. 2.5 Approaches to measuring and enhancing perceived sustainable customer value

(see Fig. 2.6); and (3) in terms of their contribution to the 17 SDGs established by the United Nations (see Fig. 2.7). If, after such an evaluation, the perceived sustainable benefits outweigh the perceived sustainable costs, the product is considered valuable or attractive from a rational consumer standpoint (See Fig. 2.8).

As illustrated in Fig. 2.8, it is essential for businesses to regularly evaluate the sustainable customer value of their brands. This ongoing assessment helps companies understand their market positioning in terms of sustainability and pinpoint opportunities for improvement. However, beyond this customer-centric evaluation, firms should also carry out a rigorous and objective life cycle assessment of the environmental, social, and economic costs and benefits associated with their sustainable production practices as presented in Chap. 4. Such assessments are critical for ensuring credibility and transparency. Without them, companies risk being accused of greenwashing—a situation that can arise when there is a significant discrepancy between a brand's actual sustainability performance and how sustainable it is perceived to be by stakeholders. Understanding and managing this gap is crucial for building trust and long-term brand equity in sustainability-driven markets. Despite its importance, relatively few methodological efforts have been made

Sustainable benefits	**Less** ◄					► **More**	
A. Before consumption *(production)*							
1. Environmental benefits	1	2	3	4	5	6	7
2. Social benefits	1	2	3	4	5	6	7
3. Economic benefits	1	2	3	4	5	6	7
B. During consumption *(purchase & usage)*							
1. Environmental benefits	1	2	3	4	5	6	7
2. Social benefits	1	2	3	4	5	6	7
3. Economic benefits	1	2	3	4	5	6	7
C. After consumption *(disposal)*							
1. Environmental benefits	1	2	3	4	5	6	7
2. Social benefits	1	2	3	4	5	6	7
3. Economic benefits	1	2	3	4	5	6	7
Sustainable costs	**Less** ◄					► **More**	
A. Before consumption *(production)*							
1. Environmental costs	1	2	3	4	5	6	7
2. Social costs	1	2	3	4	5	6	7
3. Economic costs	1	2	3	4	5	6	7
B. During consumption *(purchase & usage)*							
1. Environmental costs	1	2	3	4	5	6	7
2. Social costs	1	2	3	4	5	6	7
3. Economic costs	1	2	3	4	5	6	7
C. After consumption *(disposal)*							
1. Environmental costs	1	2	3	4	5	6	7
2. Social costs	1	2	3	4	5	6	7
3. Economic costs	1	2	3	4	5	6	7

Fig. 2.6 Research template for measuring the sustainable customer value

to systematically measure the gap between perceived and actual sustainability performance, due to the inherent complexity of such an endeavor.

One of the earliest and most notable exceptions is the collaborative study by Interbrand (2014) and Deloitte, known as the Best Global Green Brands initiative. Interbrand gathered consumer perception data to assess how the public views the sustainability of various brands. The starting point was the top 100 brands

Sustainable benefits	Less ←					→ More
1. Benefits related to SDG1	1	2	3	4	5	6	7
2. Benefits related to SDG2	1	2	3	4	5	6	7
3. Benefits related to SDG3	1	2	3	4	5	6	7
4. Benefits related to SDG4	1	2	3	4	5	6	7
5. Benefits related to SDG5	1	2	3	4	5	6	7
6. Benefits related to SDG6	1	2	3	4	5	6	7
7. Benefits related to SDG7	1	2	3	4	5	6	7
8. Benefits related to SDG8	1	2	3	4	5	6	7
9. Benefits related to SDG9	1	2	3	4	5	6	7
10. Benefits related to SDG10	1	2	3	4	5	6	7
11. Benefits related to SDG11	1	2	3	4	5	6	7
12. Benefits related to SDG12	1	2	3	4	5	6	7
13. Benefits related to SDG13	1	2	3	4	5	6	7
14. Benefits related to SDG14	1	2	3	4	5	6	7
15. Benefits related to SDG15	1	2	3	4	5	6	7
16. Benefits related to SDG16	1	2	3	4	5	6	7
17. Benefits related to SDG17	1	2	3	4	5	6	7
Sustainable costs	**Less ←**						**→ More**
1. Costs related to SDG1	1	2	3	4	5	6	7
2. Costs related to SDG2	1	2	3	4	5	6	7
3. Costs related to SDG3	1	2	3	4	5	6	7
4. Costs related to SDG4	1	2	3	4	5	6	7
5. Costs related to SDG5	1	2	3	4	5	6	7
6. Costs related to SDG6	1	2	3	4	5	6	7
7. Costs related to SDG7	1	2	3	4	5	6	7
8. Costs related to SDG8	1	2	3	4	5	6	7
9. Costs related to SDG9	1	2	3	4	5	6	7
10. Costs related to SDG10	1	2	3	4	5	6	7
11. Costs related to SDG11	1	2	3	4	5	6	7
12. Costs related to SDG12	1	2	3	4	5	6	7
13. Costs related to SDG13	1	2	3	4	5	6	7
14. Costs related to SDG14	1	2	3	4	5	6	7
15. Costs related to SDG15	1	2	3	4	5	6	7
16. Costs related to SDG16	1	2	3	4	5	6	7
17. Costs related to SDG17	1	2	3	4	5	6	7

Fig. 2.7 Measuring the sustainable customer value through the 17 SDGs

Organic jus	Consumer at the sustainability dilemma.	*Non-organic juice*

How much do you agree with the following about the sustainable value of organic juice?

Sustainable benefits						***Sustainable costs***					
1. More health benefits	1	2	3	**4**	5	1. Higher price	1	2	3	**4**	5
2. Higher nutritional value	1	2	**3**	4	5	2. Seasonal availability	1	**2**	3	4	5
3. Absence of pesticides	1	2	3	4	**5**	3. Certification skepticism	**1**	2	3	4	5
4. Lower carbon footprint	1	2	3	**4**	5	4. Taste variations	1	**2**	3	4	5
5. Fairtrade/ethical farming	1	2	3	**4**	5	5. Higher transportation costs	**1**	2	3	4	5

Note: Likert scale questions (i.e., 1 = Strongly disagree, 5 = Strongly agree).

Example:

$$\boldsymbol{Sustainable\ benefits} = \frac{4+3+5+4+4}{5} = \frac{\mathbf{20}}{\mathbf{5}} = \mathbf{4}$$

$$\boldsymbol{Sustainable\ costs} = \frac{\mathbf{4+2+1+2+1}}{\mathbf{5}} = \frac{\mathbf{10}}{\mathbf{5}} = \mathbf{2}$$

$$\boldsymbol{Sustainable\ customer\ value} = \frac{\boldsymbol{Sustainable\ benefits}}{\boldsymbol{Sustainable\ costs}} = \frac{\mathbf{10}}{\mathbf{5}} = \mathbf{2}$$

Conclusion : The organic juice is perceived as twice as valuable in terms of sustainability compared to its conventional alternative, which does not come from organic farming

Fig. 2.8 Hypothetical example for the estimation of the sustainable customer value

Sustainability perception *(Interbrand)*	**Sustainability performance** *(Deloitte)*
1. Authenticity	1. Product and services (up to 30 points)
2. Differentiation	2. Operations (up to 30 points)
3. Presence	3. Governance (up to 12.5 points)
4. Relevance	4. Stakeholder engagement (up to 12.5 points)
5. Consistency	5. Supply chain (up to 10 points)
6. Understanding	6. Transportation & logistics (up to 5 points)

Fig. 2.9 Methodology of *Best global green brands*. *Source* Own development based on Interbrand (2014)

from Interbrand's global ranking. To gather consumer insights, 15,000 participants were surveyed online. Each brand was evaluated by at least 125 respondents, resulting in approximately 1250 consumer ratings per brand. The survey was carried out in the ten largest economies by GDP, to ensure representation from the most economically influential markets. The performance score was calculated by Deloitte using 83 metrics—covering areas such as product efficiency, sustainable materials, energy and water use, emissions, waste, and supply chain environmental impacts—grouped into six weighted categories. The main group of indicators used to measure perceived and actual sustainability performance in this initiative are briefly presented in Fig. 2.9.

The comparison between sustainability perception and actual sustainability performance revealed both positive and negative perception–performance gaps. *A positive gap* occurs when a brand is perceived as less sustainable than it actually is, meaning the company's real sustainability performance exceeds public perception (e.g., Cisco and L'Oréal). Conversely, a *negative gap* arises when a brand is perceived as more sustainable than it truly is—indicating that public perception surpasses actual performance, which increases the risk of greenwashing (e.g., Coca-Cola and McDonald's). Brands showing little to no gap between perceived and actual sustainability performance were considered the *best global green brands* (e.g., Ford and Toyota). Although the study has been criticized for including only brands listed in Interbrand's best global brands—thus excluding many companies with strong sustainability performance but lower brand valuations—it still provides a valuable foundation for future research focused on measuring and managing sustainability perception gaps.

Chapter Conclusion

This chapter has introduced two interrelated yet distinct approaches to sustainability within the context of marketing. The structural approach underscores the collaborative roles of society, businesses, and policymakers in co-creating sustainable value through a shared responsibility framework. It emphasizes the systemic integration of stakeholder actions across the domains of sustainable consumption, production, and legislation. Complementing this, the dynamic, value-based approach that will

Criteria	Traditional marketing	Relationship marketing	Sustainable marketing	
			(past & now)	*(future)*
1. Origin	> 1950s	> 1980s	> 2000s	> 2030
1. Core idea	Generate profits through customer satisfaction	Generate profits by building and maintaining mutually beneficial long-term relationships with the various stakeholders.	Generate profits by contributing to social welfare and environmental preservation.	Co-create sustainable value with the various stakeholders
3. Scope	Product-oriented	Service-oriented	Communication-oriented	Performance and impact-oriented
4. Main stakeholders	Shareholders and customers	Internal stakeholders: shareholders, managers/executives, employees; External stakeholders: customers, suppliers, competitors, policy makers, NGOs, media.	Society, Business and Government	A global sustainability stakeholder network.
5. Marketing mantra	Every customer is a king.	Every relationship is unique.	Every action matters to sustainability	Every relationship builds a sustainable future.
6. Competitive strategy	Race with competition	Cooperate with competition	Respect competition	Co-create sustainable value with competition
7. Time-horizon	Short-term	Medium-term	Medium-term	Long-term

Fig. 2.10 From traditional to future-oriented sustainable marketing: a comparative overview

be further clarified in the following chapters offers a linear model for embedding sustainability into marketing practices, guiding businesses through the stages of understanding, creating, communicating, and delivering sustainable customer value. Together, these approaches form a comprehensive framework that balances collective systemic engagement with actionable marketing strategies, laying the groundwork for long-term environmental, social, and economic impact. Looking ahead, it is expected that the sustainable marketing paradigm will evolve beyond measuring perception–performance gaps toward assessing performance–impact gaps, thereby reducing the risk of misleading sustainability communications. This shift envisions the development of a global sustainability stakeholder network, where every relationship contributes to building a sustainable future (See Fig. 2.10.)

Discussion Questions and Practical Assignments

1. To what extent should consumers bear responsibility for driving sustainable development compared to businesses and policymakers?
2. Of the three stakeholder groups—society, businesses, and governments—which currently holds the most power in shaping sustainable development, and which one should ideally lead? Why? Support your answer with real-world examples from a specific industry (e.g., fashion, energy, food).
3. How can prioritizing economic sustainability (profitability) undermine environmental and social sustainability goals? Reflect on how short-term profit motives can create trade-offs with long-term planetary or societal well-being.
4. Choose a sustainable product (e.g., electric car, reusable water bottle, fair-trade chocolate). Using the sustainable customer value equation (benefits vs. costs), assess its perceived value from the consumer's perspective. Then suggest ways to either increase the perceived sustainable benefits or reduce the perceived sustainable costs.
5. Pick one of the United Nations' Sustainable Development Goals (SDGs). Design a product or service that contributes to this goal and explain how you would market it by highlighting its social, environmental, and economic impact. How would you communicate its full sustainable value to target consumers?
6. Can a product truly be considered "sustainable" if it only performs well on one pillar of the triple bottom line (e.g., high profits but poor environmental or social outcomes)?
7. Identify 2 positive and 2 negative measures adopted by policymakers in a specific country in an attempt to improve the sustainable production and consumption. What were their intended outcomes? Were they achieved: were the measures effective on a local level? Present some graphs/numbers to prove their (in)effectiveness for a certain period of time. Did these measures have global effectiveness (positive social and/or environmental impact) or some spillover effects? For example: did they inspire other countries? How many/which one(s)? Did they reduce the global CO_2 emissions? (See Table 2.1).

Table 2.1 A conceptual framework for measuring the effectiveness of sustainable legislation

Public policy measures	Local effectiveness (e.g., on a city, country regional level)			Global effectiveness (e.g., on a world level)		
	Intended outcomes	Actual outcomes		Intended outcomes (if any)	Actual outcomes	
		Positive (if any)	Negative (if any)		Positive (if any)	Negative (if any)
A. Positive measures						
1. Targeting business consumers						
2. Targeting final consumers						
B. Negative measures						
1. Targeting businesses						
2. Targeting final consumers						

References

Interbrand. (2014, June). *Best global green brands 2014: The power of participation*. https://www.slideshare.net/LobnaQassem1/interbrandbestglobalgreenbrands2014overview8pdf

Cook, J., Ellerton, P., & Kinkead, D. (2018). Deconstructing climate misinformation to identify reasoning errors. *Environmental Research Letters, 13*(2), 024018, https://doi.org/10.1088/1748-9326/aaa49f

Elkington, J. (1997). *Cannibals with forks: The triple bottom line of 21st century business*. Capstone Publishing.

Hawken, P. (2010). *The ecology of commerce: A declaration of sustainability* (Rev. ed.). Harper-Business.

United Nations World Commission on Environment and Development. (1987). *Our common future*. Oxford University Press.

Wilson, J. N. (2020, September 8). *The allure of wishful thinking*. Medium. https://medium.com/change-becomes-you/the-allure-of-wishful-thinking-c7be4b6b6831

Understanding Sustainable Value Through Consumer Behavior

3

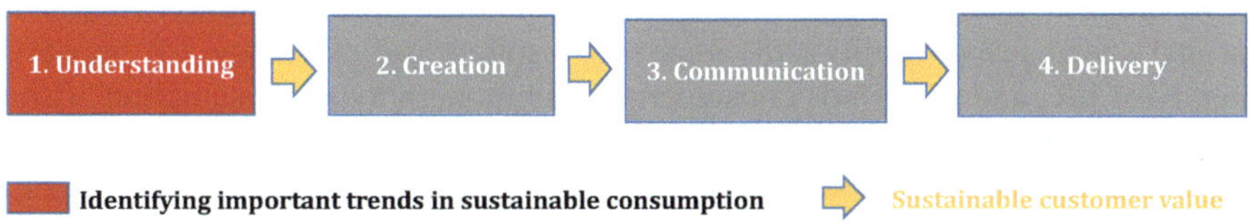

Chapter Introduction

Sustainable consumption is a concept that has gained increasing attention in recent years, as societies grapple with the challenge of balancing economic growth, environmental preservation, and social well-being. Defined as an adaptive and balanced approach to consumption that meets the needs of the present without compromising the ability of future generations to meet their own needs, sustainable consumption is both an imperative and a paradox. While the goal of a truly sustainable society seems aspirational, if not utopian, due to the inherent conflict between infinite human desires and finite planetary resources, understanding consumer behavior within this framework is crucial for creating meaningful change. This chapter explores the key dimensions of sustainable consumption, focusing on how consumers make decisions regarding what, where, when, and why they consume. It highlights the critical role of demographic, geographic, and psychographic factors in shaping sustainable consumer behaviors and examines the ways in which companies and policymakers can influence more responsible consumption patterns. Additionally, the chapter introduces the concept of "wallet voting," emphasizing the power of consumer choices in driving social and environmental impact. By framing consumption holistically—before, during, and after the act of purchasing—this discussion aims to provide a comprehensive roadmap for fostering more sustainable behaviors.

D. K. Stoyanov, *Sustainable Marketing*, Springer Texts in Business and Economics,
https://doi.org/10.1007/978-3-032-16498-8_3

3.1 Understanding Sustainable Consumption

One of the most complete definitions frames sustainable consumption "as an adaptive, balanced, and contextualized approach to consumption that (1) meets the basic needs of the current generation, (2) does not impoverish future generations, (3) does not cause irreversible damage to the environment, (4) does not create a loss of function in natural systems (ecological and human value systems; environmental and social responsibility), (5) improves resource use efficiency, (6) improves quality of life, and (7) avoids consumerism and modern hyperconsumption" (Lim, 2017, p. 71). However, our goal to become a sustainable society can also be viewed to a certain extent as a merely impossible dream or some kind of utopia because the needs of the society have always been infinite contrary to the available planetary resources. One of the pioneers in sustainable marketing even classified the term "sustainable consumption" as an oxymoron because "to 'consume' something is to use it up or to destroy it, which is the complete opposite to 'sustaining' something" (Peattie & Collins, 2009, p. 107) and therefore consumer behavior should be viewed holistically in consideration of the stages *before, during, and after consumption* (see Fig. 3.1).

Moreover, to better understand and promote sustainable consumption, marketers should ask consumers the conventional 5W questions (i.e., *Who, What, Where, When,* and *Why*)—but reformulate them in a sustainability context.

Stages of sustainable consumption	Purchase decision making process	Description	Examples
Before	1. Need recognition	Do you really need a new product? Can't you repair your existing product or borrow one?	Sharing services (sharing economy)
	2. Information search	Where and how do you look for information?	Ecosia, as an alternative search engine solution
	3. Evaluation of alternatives	Do you consider sustainability criteria when selecting a product?	Local suppliers, organic products; supporting NGO
	4. Purchase	Where will you buy your product? How will you go there? How will you pay for it?	Online/Offline, sustainable mobility
During	5. Usage	How frequently, when, and where will you use the new product?	Avoiding idling; reduce water, and electricity usage
After	6. Post-usage	What can we do with a product at the end of its lifecycle?	6Rs = reduce, re-use,. repair, repurpose, recycle and redistribute

Fig. 3.1 Scope and application of the conventional stages of sustainable consumption. *Source* Own development based on Belz & Peattie (2012)

3.1.1 Who are the Current and Potential Customers of the Product?

This question aims to identify the demographic profile of the final consumers (e.g., their age, gender, life cycle stage/marital status, occupation, religion preferences, nationality, education, and income level), as well as that of all the other players/parties that may be involved in the sustainable purchasing process (e.g., initiators, influencers, deciders, and buyers). For example, influencers play a crucial role in shaping sustainability awareness, particularly among different age groups. Younger consumers, especially millennials, often exhibit lower levels of sustainability knowledge, but when they engage with influencers who promote ethical consumption, their awareness tends to increase. Influencers serve as role models, guiding millennials toward making more sustainable choices and fostering a greater understanding of environmental issues. In contrast, older consumers show a diminished reliance on influencers for sustainability awareness. This group tends to trust their own opinions or local recommendations rather than celebrity endorsements. They often seek detailed information and transparency about products before making purchasing decisions, reflecting a more cautious approach to influencer marketing (Johnstone & Lindh, 2018).

Digital Sustainability: Cristiano Ronaldo vs Coca-Cola 1:0

The 36-year-old Portuguese football player, Cristiano Ronaldo, is one of the most successful football players of all time. Some of his international trophies and awards involve 5 Ballon d'Or, 4 European Golden Shoes, 5-time winner in Champions leagues. In 2021, he is the sport celebrity with the highest number of followers on social networks. With more than 550 million followers worldwide, he generates a large amount of his fortune from being an online influencer. According to Forbes in 2021, he is the richest football player worldwide with $125 million of total earnings, 44% of which comes from endorsements. He is well known for his healthy lifestyle and own line of products under the brand CR7.

On June 14, 2021, during one of the Euro 2020 press conferences, he took away the Coca-Cola bottles that had been left in front of him and raised a bottle of water, appealing to the audience to drink aqua (i.e., the Portuguese word for water). This rebel reaction/gesture toward one of the long-standing sponsors in the history of global sport events (e.g., football championships and Olympic games) lasted for around 10 s but had/evoked the following important domino and butterfly effects:

- The Italian football player, Manuel Locatelli, also replaced the bottles during a press conference with a bottle of water, therefore joining the "football club" against fizzy drinks

- The French football player, Paul Pogba, a devout Muslim, also followed the example of Cristiano Ronaldo by removing a bottle of Heineken placed in front of him during a press conference (without knowing that the beer was actually with 0% alcohol)
- The Swedish retailer Ikea launched in Canada a reusable glass water bottle named "Cristiano" and a related social media campaign under the slogan "*Drink water sustainably*"
- The short video from the press conference got viral on social networks, being shared by millions of online users: individuals, companies, and NGOs. For example, the World Health Organization (WHO) tweeted "Stay fit & healthy like @Cristiano by 💪reducing sugar consumption and limiting intake of sweets & sugary drinks." The French mineral water brand Evian posted the video under the title: "Couldn't have said it better ourselves!"

Source: Based on different public sources

3.1.2 What do Consumers do Before/During/After the Product Consumption?

Some of the typical questions that consumers can be asked are the following ones:

- What kind of sustainable foods do they purchase (e.g., fresh, unprocessed, organic, locally grown, seasonal, or animal-based products), in what quantities and at what frequency?
- What kind of transport do they use for short and long distances, to go home, to work, or on a vacation (e.g., airplane, private car, car sharing, ride sharing, P2P car rental, public transport, bicycle, rollers, skateboard, or walking) and at what frequency?
- What do they do with the main product, its related accessories, and packaging at the end of their life cycle (e.g., dispose, repair, reuse, recycle, resell, borrow, or donate them)?
- What sources of information do they use at each stage of the consumption process regarding the purchase, usage, and disposal of the product (e.g., friends and relatives, mass media, social networks, corporate, governmental, or NGO platforms and applications) and what is their level of importance in the decision-making process?

Digital sustainability: The Case of Yuka

Yuka is a French mobile application available for free on both Android and IOS, that allows consumers to scan the barcodes of 1,5 million food and 500 thousand cosmetic products in order to receive a detailed health-related information on them. It was launched in 2017 with the objective to help from one side the society to improve its consumption habits (e.g., by raising the awareness of the consumers to the negative impact that certain ingredients can have on their health) and from another one—to encourage companies to constantly replace the potentially dangerous ingredients in their products with healthier alternatives. The health-related risk of product consumption is measured on a 100-point scale through four categories (i.e., no risk, low risk, moderate risk and high risk), and there is a detailed explanation for every product on why the application provided a particular score. When a product receives a low score (e.g., it is associated with a moderate or high risk) the users receive recommendations on the healthier alternatives that they can purchase instead. In March 2021, the company reported having 21 million users in 12 countries (i.e., France, Belgium, Switzerland, Luxembourg, Spain, Italy, Germany, United Kingdom, Ireland, USA, Canada, and Australia).

The mobile platform claims being 100% independent from third-party organizations and having as a sole objective to inform the consumers on how to improve their consumption habits. It has the following three guiding principles: (1) not to receive any funding from brands and manufacturers, and not to allow anyone to influence their evaluations. (2) not to display any advertising in the application. (3) to respect data privacy by neither exploiting nor selling any user data. Its business model is based on three main sources of revenues: (1) the premium version of the application that provides users with additional features such as a search bar, an offline mode, and customizable alerts; (2) a guidebook on healthy eating that suggests different recipes and provides some tips on health nutrition; (3) a seasonal food calendar that suggests which fruit and vegetables to be consumed according to the season.

Source: Based on different public sources

3.1.3 Where do Customers Purchase/Use/Dispose the Products?

Through this question marketers and policymakers want to understand the geographic characteristics of the consumers (e.g., urban vs suburban vs rural population, country, neighborhood). Some of the typical sub-questions are the following ones:

- Where do they purchase the product (e.g., online vs offline, from local farmers vs national/global retail chains, slow vs fast fashion retailers, food banks vs regular stores)?
- Where do they use the product (e.g., at home, at work, on the street, or in the garden)?
- Where do they dispose their product waste (e.g., on the ground, into water bodies, in the landfill, general waste, or recycling bins)?
- Where do they live (e.g., eco-friendly vs energy-efficient vs smart homes)?

Imagine buying an apple. If the apple was grown locally, it requires less transportation, which reduces fuel use and emissions. However, if the same apple was shipped from a farm 1000 km away, the environmental impact increases because of the fuel burned during transport. By asking "Where was this apple grown?", a consumer can make a more sustainable choice, supporting local agriculture and reducing their carbon footprint.

3.1.4 When do Customers Look for/purchase/use/dispose the Products?

This question enables marketers and policymakers to draw the three-stage consumer timeline before, during, and after the consumption. Some typical sub-questions include the following ones:

- When do they buy products (e.g., sales periods, holidays, or when they need them)?
- When and for how long do they use their most energy consuming devices (e.g., peak vs off-peak electricity hours)?
- When are they more willing to donate money, clothes, or any other products for charity (e.g., around important religious holidays, at no special occasion, or never)?

Digital Sustainability: The Case of Too Good to Go

Too Good To Go (TGTG) is a Danish mobile application created in 2015 with the mission to fight against the food waste generated worldwide. It is available for free download on App Store and Google play and provides a platform where food establishments such as restaurants, bakeries, and convenience stores can offer consumers their unsold surpluses (e.g., a set of products with a close expiration date) at lower/preferential prices in the form of baskets to save. Thus, retailers can reduce the social and environmental impacts related to food waste and still generate certain profits. As for 2021, Too Good To Go reports having 46.3 million users in 15 countries and 96.9 meals saved. According to Statista, it is the most famous anti-food waste app in France (2019), the most frequently used sustainable app in the UK (2021) and the most frequently downloaded free mobile app by Polish consumers in September 2021.

Source: Based on different public sources

3.1.5 Why do Customers Purchase/Use/Dispose the Products?

This question targets the identification of the psychographic characteristics of the consumers—their personalities (extroverted vs introverted; thinking vs feeling; sensing vs intuitive; judging vs perceptive), hobbies and lifestyles, self-image (positive vs negative), and their prevailing needs and motivations. Some of the relevant sub-questions that marketing researchers can ask consumers are as follows:

- Why do current customers look for/buy/use/dispose/recycle the products of the company (e.g., product extension vs product depletion, hedonistic vs ascetic motivations)?
- Why do potential customers not look for/buy/use the products of the company?

"Why" is often considered the most essential question in marketing research because it helps uncover the underlying drivers of consumer needs—the deep psychological triggers of human behavior. Traditional marketing typically relies on Maslow's (1943) hierarchy of needs, which classifies human motivations into five ego-oriented categories: physiological, safety, belonging, esteem, and self-actualization. However, we argue that this widely used framework does not fully account for the motivations that are particularly salient in a sustainability context. To address this gap, we propose an additional sixth level—*other-oriented need*—which captures motivations such as helping others, contributing to community well-being, engaging in charitable actions, and supporting broader societal improvement. Following the logic of the original hierarchical model, this sixth level typically emerges once lower-level needs are sufficiently satisfied, aligning with the model's assumption of sequential need fulfillment. Although this assumption is not universally valid—altruistic individuals may prioritize others' well-being even before meeting their own—the expanded model offers a useful perspective for interpreting differences in sustainability engagement across contexts. It helps clarify, for example, why consumers in developing countries, who are often focused on meeting basic needs, generally display lower sustainability involvement, whereas consumers in highly developed economies may be more inclined toward prosocial and sustainability-oriented behaviors.

Lim (2017) suggested an integrated framework of sustainable consumption which involved:

(1) *A sustainability mindset*: a sense of care for nature, for your own, and the community which results in a social, ethical, and environmental consciousness and responsibilities (i.e., *responsible consumption*). For example, consumers are not simply concerned about sustainability, but they are aware of the potential consequences of their behavior, and the fact that with every purchase, they can make a social and environmental contribution (i.e., a term to which we refer as "wallet voting"—using your purchasing power to make the difference you want to see in the world).
(2) *Sustainability practices*: they involve the adoption of various anti-consumption attitudes and behaviors (i.e., *aversion, avoidance, and abandonment*) related to the partial or complete resistance to the consumption and/or marketing of products (e.g., downshifting, ad blocking, and anti-spam techniques). *Aversion* means having an extremely negative attitude toward a particular topic (e.g., meat consumption) or subject (e.g., product, brand, or a company). It can trigger moderation in the public and private product consumption (e.g., responsible water consumption in countries where water is a free commodity, purchasing high-tech products, clothes, and luxury items at lower frequency and not simply because it is trendy). Avoidance and aversion both represent behavioral consumer reactions. However, avoidance reflects the behavior of consumers that have never used certain products (e.g., guns, drugs, or cigarettes), and abandonment—quitting the consumption of previously consumed desirable or undesirable products for objective reasons (e.g., due to food allergies, addictions, health problems, or lack of resources) or subjective reasons (e.g., aversion evoking strong negative feelings). Abandonment, as an anti-consumption mode, has three forms: *rejection* (e.g., vegans and vegetarians), *restriction* (e.g., flexitarians), and *reclamation.*

3.2 Sustainable Gaps in Consumer Behavior: The Roadmap to Sustainability

Sustainable gaps in consumer behavior can be defined as all types of cognitive, affective, conative, and behavioral differences/barriers existing in the sustainability continuum between ignorance and impacts. While consumer behavior is not always as sequential and straightforward as suggested by our model, and some strategies can address multiple gaps, this framework can still serve as a strategic tool for businesses and policymakers. It can help identify the most significant sustainability gaps within a particular society and provide guidance on how to address them effectively (see Fig. 3.2). Additionally, individuals can use this model as a self-assessment tool to better understand and improve their consumption habits.

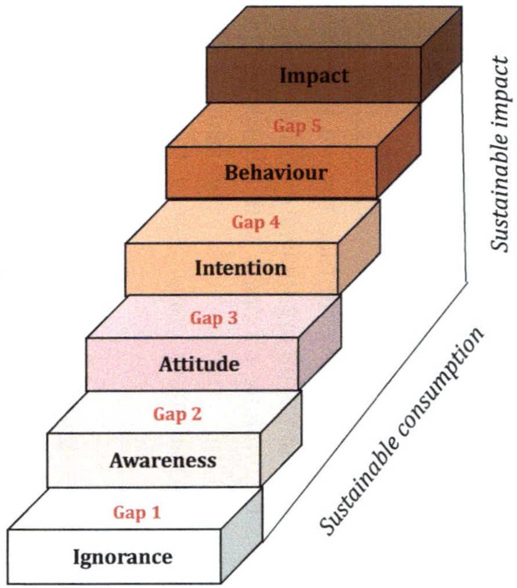

Sustainable gaps	Current state	Future state	Strategy
1	Ignorance	Awareness	Educational campaigns, influencer marketing
2	Awareness	Attitude	Emotional advertising
3	Attitude	Intention	Providing tangible and intangible incentives
4	Intention	Behviour	Removing action barriers introducing sanctions
5	Behaviour	Impact	Transparency and tracking footprint calculators.

Fig. 3.2 The roadmap to sustainability: a strategic analytical tool

3.2.1 Ignorance vs Awareness

This gap arises because consumers may not recognize the sustainability issues, such as the damage caused by plastic waste, until they are exposed to relevant information. The shift from ignorance to awareness marks the transition from being uninformed to becoming conscious of sustainability challenges. For example, consumers who are unaware of the environmental impact of plastic waste may regularly purchase products with excessive plastic packaging. Once they are educated about the environmental consequences (e.g., pollution, wildlife harm), they may become more conscious of their consumption habits. To address this gap, educational campaigns and clear, engaging information are crucial. Governments and organizations can collaborate with media outlets, influencers, and NGOs to highlight the adverse effects of plastic waste. Visual storytelling, such as documentaries and social media content, can be effective in raising awareness. Retailers can also use product labeling to inform consumers about sustainable alternatives, and school curriculums can integrate sustainability topics to educate younger generations early on. By making sustainability education widespread and accessible, consumers can transition from ignorance to awareness more effectively. According to a recent article published in one of the most reputable and influential peer-reviewed medical journals in the world, *The Lancet*, plastics pose a serious and increasingly underestimated threat to both human health and the environment. They contribute to illness and premature mortality across all age groups and generate annual health-related economic losses exceeding US$1.5 trillion, disproportionately affecting vulnerable and low-income populations. This crisis is primarily driven by the rapid increase in plastic production, which has risen from 2 megatons (Mt) in 1950 to 475 Mt in 2022, with projections reaching 1,200 Mt by 2060. Environmental pollution has grown in parallel, with approximately 8,000 Mt of plastic waste now contaminating the planet, of which less than 10% is recycled (Landrigan et al. 2025).

3.2.2 Awareness vs Attitudes

This gap exists when consumers are aware of the issue (e.g., the harmful effects of fast fashion) but do not develop a negative attitude toward it or prioritize sustainability in their decisions. Consumers who are unaware of the environmental impact of plastic waste may regularly purchase products with excessive plastic packaging. Once they are educated about the environmental consequences (e.g., pollution, wildlife harm), they may become more conscious of their consumption habits. Consumers who are unaware of the environmental impact of plastic waste may regularly purchase products with excessive plastic packaging. Once they are educated about the environmental consequences (e.g., pollution, wildlife harm), they may become more conscious of their consumption habits. Bridging this gap requires emotional engagement and value alignment. Campaigns should focus on fostering empathy and personal connection to environmental issues, such as showcasing the suffering of wildlife affected by plastic pollution or illustrating the impact on future generations. Social norm interventions, like emphasizing how many others have adopted sustainable behaviors, can also influence attitudes. Additionally, providing positive role models and reinforcing the social and moral value of sustainable practices can help consumers internalize and adopt pro-environmental attitudes.

Carol Dweck (2006), a professor of psychology at Stanford university classified people into two groups: (1) *People with a fixed mindset*: they are often characterized by a high intellectual capacity (IQ) but a narrow and deterministic view of the world. They consider that they are who they are, and they cannot change which can be problematic if they face serious challenges in life or they have to adopt a significant behavioral shift. (2) *People with a growth mindset:* they have the attitude of open-mindedness that enable them to easily leave their comfort zones and embrace changes more smoothly, as they perceive them as growth opportunities rather than challenges. Though this classification was not established specifically in a sustainability context it is useful in demonstrating how being intelligent can be less important that having the right attitude, or in other words that being aware on a particular topic may not necessarily translate into a specific attitude toward it. Similarly, there are nowadays some well-educated, highly intelligent, and very powerful people that are fully conscious on the gravity of many sustainability problems, but still do not have the perceived sense of urgency when it comes to taking some responsible actions. The gap between awareness and attitude is basically a gap between knowledge and emotional engagement (e.g., yes, I know but I do not care).

3.2.3 Attitudes vs Intentions

The attitude-intention gap arises when consumers hold positive beliefs about sustainability but fail to form strong intentions to act on these beliefs due to barriers such as cost, convenience, or availability. Consumers may have a positive attitude toward buying sustainably produced goods (e.g., organic food, eco-friendly products), but they do not necessarily intend to purchase them because they perceive them as too expensive or not widely available. Being concerned about society and the environment but not having the intention to change is also a recurrent gap in sustainable consumption. For example, consumers can be really excited about the idea of breathing pure air and living in a greener environment, but this may not necessarily reflect their intention to take certain actions in this direction (e.g., I am really concerned about the environment and I really like the overall idea of recycling but as I live in a small apartment I just do not have enough space to separate paper from plastic, glass, and aluminum containers).

3.2.4 Intentions vs Actions

This gap occurs when consumers have the intention to engage in sustainable behavior (e.g., reducing personal emissions) but fail to translate that intention into actual action due to competing priorities, convenience, or habits. As the English philosopher John Locke stated, *actions are the best interpreters of men's thoughts* (Locke, 1690/1975). A consumer may intend to reduce their carbon footprint by driving less or using public transport more frequently. However, when it comes time to choose between driving and using public transportation, the convenience and comfort of driving may lead to the consumer opting for the car. Bridging the gap between positive attitudes and actual intentions to act sustainably requires addressing perceived barriers such as cost, accessibility, and convenience. For example, making sustainable products more affordable through subsidies, discounts, or bulk-purchase incentives can encourage consumers to align their attitudes with their purchase intentions. Educating consumers about the long-term cost savings or health benefits of eco-friendly choices can also help. Moreover, enhancing convenience, such as providing more accessible recycling points or offering compact, home-friendly waste-separation solutions, can motivate action. Campaigns that emphasize the alignment of personal values with societal and environmental benefits may also foster stronger intentions by making sustainable actions feel achievable and rewarding.

The classic theory of planned behavior (Ajzen, 1991) can be extremely useful in understanding this specific gap in sustainable consumption because it assumes that an individual's behavior is shaped by their intentions, which are influenced by three key factors:

(1) *Attitude toward the behavior*: this refers to the individual's overall evaluation of a specific behavior, i.e., whether they perceive it as positive or negative. For example, if a consumer believes that recycling or purchasing eco-friendly products benefits the environment, they are more likely to engage in these behaviors.
(2) *Subjective norms*: this factor reflects the perceived social pressure to engage or not engage in a particular behavior. It includes the individual's perception of how others in their social network (e.g., family, friends, colleagues) view the behavior. For example, if an individual's peers strongly support sustainable behaviors such as reducing plastic waste, the individual may feel compelled to adopt similar behaviors due to social pressure.
(3) *Perceived behavioral control*: this refers to an individual's perception of how easy or difficult it is to perform a particular behavior, which is influenced by factors such as past experiences and anticipated obstacles. For instance, someone might want to recycle but feel discouraged if they believe the recycling system is complicated or that there are limited facilities for waste sorting in their area. In this case, despite having the intention to recycle, the individual's behavior might be hindered by perceived barriers.

These three factors—attitudes, subjective norms, and perceived behavioral control—together shape an individual's intentions, which in turn guide their actual behavior. However, external factors such as economic constraints or resource availability can further modify the likelihood of the behavior occurring. For instance, in countries like Canada, where consumers pay a fixed monthly fee for water, there is no direct cost related to water usage, leading to higher consumption despite intentions to conserve. Similarly, in France, where employees of EDF receive free electricity, the lack of financial consequences may lead to higher energy consumption, despite the intention to use energy efficiently.

Given these dynamics, policymakers and NGOs can use the insights from the theory of planned behavior to encourage more responsible consumer behavior. For example, during the COVID-19 pandemic, governments could have employed the following strategies to increase vaccination rates and mask-wearing compliance: (1) changing individuals' attitudes toward vaccination (e.g., providing more information about its benefits), (2) influencing perceptions of societal norms (e.g., highlighting that peers, family, and society are supportive of vaccination), and (3) altering individuals' perceptions of the barriers to vaccination or mask-wearing (e.g., addressing concerns about side effects or discomfort). These strategies, rooted in the theory, can help shape behavior by addressing the cognitive, social, and control-related factors that influence consumer actions.

3.2.5 Actions vs Impacts

This gap exists when consumers engage in sustainable behaviors, but those actions may not lead to significant positive environmental outcomes due to factors beyond the individual's control, such as insufficient infrastructure or the limited effectiveness of sustainable alternatives. A consumer may consistently engage in sustainable behaviors such as recycling, using reusable bags, and purchasing eco-friendly products. However, their overall impact on the environment may still be limited due to larger systemic issues, such as the environmental impact of the manufacturing processes behind "eco-friendly" products, or the inefficiencies in local recycling systems. To close the gap between individual actions and their tangible environmental impacts, systemic improvements and greater transparency are crucial. Governments and organizations can invest in more efficient recycling systems and infrastructure to ensure that consumers' recycling efforts are not wasted. Similarly, businesses should improve the sustainability of their supply chains, reducing the environmental footprint of manufacturing processes for "eco-friendly" products. Certifications and clear, verified labels can help consumers make more impactful choices by distinguishing genuinely sustainable products. Additionally, fostering collaboration between stakeholders—governments, businesses, and consumers—can help align individual efforts with systemic change, ensuring that actions translate into meaningful environmental outcomes.

This gap is closely linked to the concept of "unbalanced consumption," which refers to the unequal distribution of economic wealth among nations. This inequality leads to significantly higher levels of consumption and material footprints in developed and emerging economies, while developing countries have comparatively low levels of consumption (see Fig. 3.3). Sustainable consumption, however, is often assessed based on the types of products people buy or consume, rather than considering the resources they abstain from using. As Beltz and Peattie (2012) highlight, a consumer who makes numerous prosocial and environmentally friendly purchases may be classified as sustainable, while someone who lives frugally out of necessity rather than choice is often overlooked. For instance, an individual in a high-income country consumes an average of 28 tons of resources per year, compared to less than 2 tons for someone in a low-income nation. If everyone consumed at the level of a typical high-income country resident, we would require the resources of 3.8 Earths to sustain such demand. Conversely, aligning global consumption patterns with those of lower-income nations could establish a more sustainable balance (Hickel, 2017). For example, China was the largest emitter of carbon dioxide (CO_2) in 2023, accounting for over 31% of global emissions. Along with the United States, India, Russia, and Japan, it ranked among the top five polluters, collectively contributing approximately 60% of worldwide CO_2 emissions (Statista, 2024). These nations are now focusing on cultivating sustainability-oriented mindsets, raising awareness about social and environmental challenges, and incentivizing behavioral shifts. While these efforts may appear to stem from a newfound sense of moral duty, they are also essential for safeguarding their long-term economic and social stability. Without a commitment to responsibility, their power and influence are at risk. As such, addressing the root causes of unbalanced consumption and ensuring global equity in sustainability efforts are crucial.

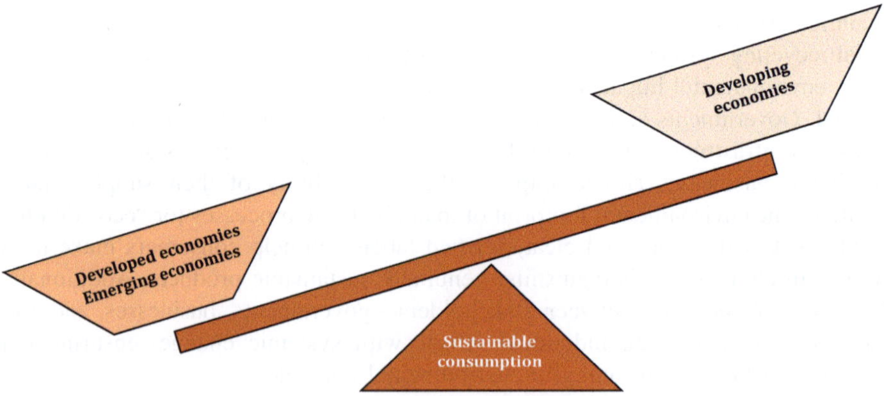

Fig. 3.3 The concept of unbalanced consumption

3.3 Psychological Effects in Sustainable Consumption

Understanding psychological phenomena that influence consumer behavior is essential to addressing sustainability gaps. These biases shape how individuals make decisions and their impact on sustainable consumption. Here are six key psychological effects:

(1) **self-licensing effect**—it is the subconscious belief that performing a good deed in the past justifies less responsible actions in the future. The self-licensing effect was first introduced by Monin and Miller (2001) when studying how performing moral actions could give individuals a "license" to act less morally afterward. Sustainable consumers often strive to balance their good and bad deeds—a natural struggle, as perfection is unattainable. For example, if someone from Hawaii needs to study in New York, they have little choice but to take a plane. To offset their carbon footprint, they might calculate emissions and contribute to environmental projects using carbon offset calculators. However, self-licensing reflects a different mindset. For instance, consumers who replace traditional light bulbs with energy-efficient ones may feel justified in leaving the lights on longer, potentially increasing their overall energy use. Another example is someone who participates in an Earth Day parade but afterward indulges in an environmentally unsustainable habit, like eating a Big Mac. A study by Mazar and Zhong (2010) even found that consumers were more willing to cheat or steal after purchasing green products, indicating that the simple act of "going green" doesn't inherently make someone more ethical. Self-licensing is particularly prevalent among those with low sustainability commitment, who still view sustainable actions as a cost–benefit tradeoff rather than an ingrained mindset. This underscores the importance of incremental changes toward sustainability to avoid moral licensing, which can lead to larger ethical compromises and exacerbate sustainability gaps.

(2) **cognitive dissonance**—it occurs when consumers experience conflict between their beliefs, values, or feelings and their actual behavior (Festinger, 1957). This psychological discomfort often arises in sustainability contexts. For example, a consumer may fully understand the importance of buying organic products but cannot afford them. This conflict between values and financial limitations creates dissonance. Another example involves waste management: after China stopped importing plastic waste in 2018, much of Europe's plastic waste was redirected to Malaysia, where inadequate facilities led to harmful disposal practices, including burning plastics in landfills. These actions caused severe respiratory issues for locals. Despite being aware of such inefficiencies in waste management, some French consumers might continue separating their waste, motivated by the hope that government policies will eventually improve. This demonstrates how cognitive dissonance can persist even when individuals feel disillusioned with the broader system. A recent study conducted by the Agence nationale de sécurité sanitaire de l'alimentation (2025),

the French health agency, revealed that beverages in glass bottles contain significantly higher levels of microplastics compared to those in plastic bottles, cans, or cartons. The research examined a variety of drinks, including water, soda, iced tea, wine, and beer, to assess the extent of microplastic contamination and the influence of packaging materials. The study found that beverages in glass bottles contained an average of approximately 100 microplastic particles per liter, whereas those in plastic bottles or cans had 5 to 50 times fewer particles. Notably, water and wine exhibited relatively low microplastic levels across all packaging types. The primary source of contamination in glass-bottled beverages was identified as the paint on the metal caps, which sheds microplastic particles due to friction during storage and handling. These findings can create cognitive dissonance for consumers who prioritize sustainability. Many sustainable consumers choose glass over plastic, believing it is better for both the environment and personal health. Learning that glass bottles may actually expose them to higher microplastic contamination challenges these assumptions and generates an uncomfortable conflict between their values and behavior. To resolve this tension, consumers may feel compelled to change their purchasing habits, rationalize their choices, or seek further information about alternative sustainable packaging, highlighting the psychological complexity behind environmentally conscious decision-making.

(3) **halo effect—**it describes how individuals attempt to create a favorable impression by exaggerating their sustainable behaviors, especially during research or social interactions. It was first observed by Thorndike (1920) who found that people's overall impression of an individual could influence unrelated judgments about that person's character or capabilities. The concept has been extended to consumer behavior, particularly in sustainability research. For example, a participant in a survey might claim they regularly recycle and buy eco-friendly products to appear responsible, even if their actual behavior is inconsistent. This bias can distort research findings and make it difficult to identify genuine consumer behavior, creating a misleading picture of sustainability adoption.

(4) **mere-measurement effect**—it occurs when asking consumers about their sustainable habits prompts them to adopt better behaviors, even if the intention wasn't present before. This psychological phenomenon has been extensively documented in consumer research (Morwitz & Fitzsimons, 2004). For instance, if a consumer is asked in a survey how often they bring reusable bags to the store, they might start using them more frequently after realizing the importance of the behavior. Repeated exposure to such questions can create a cumulative effect, encouraging moderation in consumption and heightened awareness of sustainability issues. In other words, the simple act of measuring (asking consumers the right questions) can initiate a behavioral change. However, the effect may be weaker when respondents can identify the organization funding the study, as they may suspect a hidden agenda; thus, transparency and trust are essential for sustaining its positive impact. Extending this perspective, it would be valuable to investigate whether the mere-measurement effect also

occurs when consumers are repeatedly asked about the environmental costs of technologies such as generative AI systems, which impose substantial impacts not only through electricity consumption but also via carbon emissions, water use for cooling, and the resource-intensive production of hardware. Although the exact energy required to train and operate a model such as OpenAI's GPT-3 is challenging to measure. In a 2021 study, researchers from Google and the University of California, Berkeley, calculated that training the model required approximately 1,287 megawatt hours of electricity—sufficient to supply power to roughly 120 typical U.S. households for one year—and resulted in the emission of around 552 tons of carbon dioxide (Zewe, 2025). As AI adoption continues to grow worldwide, only time will reveal whether repeatedly asking consumers about their willingness to use AI tools despite associated trade-offs can trigger a mere-measurement effect and foster greater behavioral change.

(5) **bandwagon effect**—it occurs when individuals are influenced by the actions of others, often adopting behaviors because they perceive them as popular. The term originated from the study of political science, particularly in Hyman's (1942) work on how social influence impacts voting behavior. It was derived from the phrase "jumping on the bandwagon." For example, a restaurant that displays a sign saying, "90% of our customers bring their own reusable containers," can motivate other patrons to follow suit. Social cues like this tap into the human desire to conform, helping to bridge sustainability gaps by leveraging collective influence.

(6) **confirmation effect**—refers to the tendency to seek out or interpret information that aligns with pre-existing beliefs while ignoring contradictory evidence. This effect was introduced by Wason (1960) during his experiments on hypothesis testing when he demonstrated how people tend to seek information that supports their existing beliefs, which has since been extensively explored in decision-making, including in the context of sustainability. For example, a consumer skeptical of electric vehicles (EVs) might focus on articles about the environmental costs of EV battery production, overlooking studies that highlight the significantly lower lifetime emissions of EVs compared to gas-powered cars. This selective attention reinforces biases, preventing individuals from fully embracing sustainable options and perpetuating gaps in sustainability understanding.

In conclusion, understanding psychological phenomena such as self-licensing, cognitive dissonance, the halo effect, the mere-measurement effect, the bandwagon effect, and confirmation bias is crucial in addressing sustainability gaps in consumer behavior. These biases can significantly influence how consumers perceive and act upon their sustainability-related choices, often leading to contradictions between their intentions and actions. Recognizing these psychological effects can help in designing more effective strategies for promoting sustainable consumption, encouraging individuals to take small, consistent steps toward aligning their behaviors with their environmental values.

Chapter Conclusion

Achieving sustainable consumption requires a fundamental shift in both individual mindsets and systemic structures. As explored in this chapter, consumer behavior is influenced by various psychological, social, and economic factors that create gaps between awareness, attitudes, intentions, actions, and actual impact. While knowledge and education play a critical role in shaping sustainable choices, emotional engagement and structural support mechanisms are equally necessary to bridge these gaps effectively. Sustainability is not solely about promoting eco-friendly products or reducing waste—it is about cultivating a broader consciousness of how consumption impacts the planet and future generations. Businesses, policymakers, and consumers must work collaboratively to drive sustainable behavior by addressing economic and psychological barriers, enhancing transparency, and fostering a culture of mindful consumption. By embracing a holistic approach that integrates sustainability mindsets and practices, societies can move closer to narrowing the gap between aspiration and reality in achieving a more sustainable future.

Discussion Questions and Practical Assignments

1. Compulsive purchasing behavior, also known as "oniomania" (from Greek *onios* means "for sale," and *mania—insanity)* is a mental disorder associated with an obsessive and uncontrollable shopping desire and resulting in serious psychological, behavioral, and financial consequences for the suffering individual. Look on the internet for some recent studies and/or testimonials on compulsive purchasing behavior and try to respond to the following questions.

   Which human need(s) does compulsive shopping behavior satisfy and why? Why is it often classified as a mental disorder with severe consequences? Provide some specific examples of undesirable consequences caused by excessive shopping (e.g., consequences to the individual, society, and the environment). How is it different from other common types of addictions (e.g., sex, work, food, drugs, alcohol, and gambling)? In what way is compulsive shopping in the digital world different from the one in the real world? Are you familiar with some strategies or techniques that companies use to make consumers more addicted to shopping? What positive and negative measures should policymakers adopt in the fight against compulsive shopping? Suppose that you have a friend addictive to shopping, what would you advise him/her?
2. Deshopping is defined as "the deliberate – and arguably inappropriate – return of goods for reasons other than actual faults in the product. In its pure form, it is premeditated prior to and during the consumption experience as at least a potential outcome of the event" (Schmidt et al., 1999, p. 292). What motivates individuals to engage in deshopping, such as the perception of adopting a "Robin Hood" behavior and providing social justice? Which product categories are most susceptible to deshopping, and why? Do you agree that "[d]eshopping is like virtual sex, like alcohol without the hangover, like food without nutritional value" (Schmidt et al., 1999, p. 292)? Can deshopping be categorized as a sustainable

consumer behavior? Why or why not? What are the potential consequences of deshopping for consumers, retailers, society, and the environment? How can consumers be holistically educated about all these consequences? Are there ways for retailers to communicate their policies more effectively to either completely discourage deshopping or, at the very least, minimize its impact? Is this effort truly necessary?

3. How do growth and fixed mindsets influence consumer choices regarding sustainable products? Reflect on how marketers might tailor campaigns to address these different attitudes.
4. To what extent do you believe influencer marketing can genuinely shift sustainability awareness and practices among various demographic groups? Provide examples from recent campaigns or notable public figures.
5. In teams, design a new product or service that addresses one of the United Nations' SDGs. Use the sustainable customer value equation (benefits vs. costs) to demonstrate its appeal. Prepare a short marketing pitch that highlights its social, environmental, and economic impact. Include strategies for communicating its sustainable value to diverse consumer segments.
6. Discuss the psychological effects in sustainable consumption, focusing on how cognitive biases and behavioral mechanisms shape consumer decisions. In your answer, illustrate the self-licensing effect with an example and suggest how marketers can discourage moral licensing, explain how cognitive dissonance appears in sustainable purchasing and how it may be resolved through marketing or policy interventions, and evaluate how the bandwagon effect and social proof can accelerate adoption of sustainable products. Finally, reflect on the role of confirmation bias in hindering sustainable choices and propose strategies marketers can use to overcome entrenched beliefs.
7. Critically assess the concept of *unbalanced consumption* by examining how global wealth disparities shape sustainable consumption patterns and their environmental impacts. Select one high-income country (e.g., United States, Germany) and one low-income country (e.g., India, Nigeria), and use publicly available data (e.g., Statista, UN reports) to compare their material footprint per capita and CO_2 emissions. In a short essay, analyze the social, environmental, and economic consequences of these differences, and propose marketing strategies tailored to each country's specific context and sustainability challenges.
8. Analyze how digital platforms (e.g., Yuka, Too Good To Go, Ecosia, or others) influence consumer behavior and promote sustainable choices. Consider how transparency and independence build user trust, and how the *mere-measurement effect* can be applied in digital marketing to encourage responsible consumption. Select one platform and conduct a comprehensive review of its design, transparency, revenue model, and impact on user behavior. Present your findings in a report, including recommendations for improving effectiveness, engagement, and the platform's role in fostering sustainable consumption.

References

Agence nationale de sécurité sanitaire de l'alimentation. (2025, June 20). Les capsules des bouteilles en verre contaminent les boissons avec des microplastiques. [The caps of glass bottles contaminate beverages with microplastics] https://www.anses.fr/fr/content/les-capsules-des-bouteilles-en-verre-contaminent-les-boissons-avec-des-microplastiques

Ajzen, I. (1991). The theory of planned behavior. *Organizational Behavior and Human Decision Processes, 50*(2), 179–211. https://doi.org/10.1016/0749-5978(91)90020-T

Belz, F. M., & Peattie, K. (2012). *Sustainability marketing* (2nd ed.). Wiley & Sons.

Dweck, C. S. (2006). *Mindset: The new psychology of success.* Random House Digital, Inc.

Festinger, L. (1957). *A theory of cognitive dissonance.* Stanford University Press.

Hickel, J. (2017). *The divide: A brief guide to global inequality and its solutions.* Random House.

Hyman, H. H. (1942). The psychology of the bandwagon. *Social Research, 9*, 92–94.

Johnstone, L., & Lindh, C. (2018). The sustainability-age dilemma: A theory of (un) planned behaviour via influencers. *Journal of Consumer Behaviour, 17*(1), e127–e139. https://doi.org/10.1002/cb.1693

Landrigan, P. J., Dunlop, S., Treskova, M., Raps, H., Symeonides, C., Muncke, J., Spring, M., Stegeman, J., Almroth, B. C., Chiles, T. C., Cropper, M., Deeney, M., Fuller, L., Geyer, R., Karasik, R., Mafira, T., Mangwiro, A., Matias, D. M.,... & Rocklöv, J. (2025). The lancet countdown on health and plastics. *The Lancet, 406*(10507), 1044–1062.

Lim, W. M. (2017). Inside the sustainable consumption theoretical toolbox: Critical concepts for sustainability, consumption, and marketing. *Journal of Business Research, 78*, 69–80. https://doi.org/10.1016/j.jbusres.2017.05.001

Locke, J. (1690/1975). *An essay concerning human understanding* (P. H. Nidditch, Ed.). Clarendon Press.

Morwitz, V. G., & Fitzsimons, G. J. (2004). The mere-measurement effect: Why does measuring intentions change actual behavior? Journal of Consumer Psychology, 14(1–2), 64–74. https://doi.org/10.1207/s15327663jcp1401&2_8

Maslow, A. H. (1943). A theory of human motivation. *Psychological Review, 50*(4), 370–396. https://doi.org/10.1037/h0054346

Mazar, N., & Zhong, C. B. (2010). Do green products make us better people? *Psychological Science, 21*(4), 494–498.

Monin, B., & Miller, D. T. (2001). Moral credentials and the expression of prejudice. *Personality and Social Psychology Bulletin, 27*(5), 700–711.

Peattie, K., & Collins, A. (2009). Guest editorial: Perspectives on sustainable consumption. *International Journal of Consumer Studies, 33*(2), 107–112. https://doi.org/10.1111/j.1470-6431.2009.00758.x

Schmidt, R. A., Sturrock, F., Ward, P., & Lea-Greenwood, G. (1999). Deshopping– the art of illicit consumption. *International Journal of Retail & Distribution Management, 27*(8), 290–301. https://doi.org/10.1108/09590559910288569

Statista. (2024, November 13). Distribution of carbon dioxide emissions worldwide in 2023, by select country [Graph]. *Statista.* Retrieved December 4, 2024, from https://www-statista-com.scd-rproxy.u-strasbg.fr/statistics/271748/the-largest-emitters-of-co2-in-the-world/

Thorndike, E. L. (1920). A constant error in psychological ratings. *Journal of Applied Psychology, 4*(1), 25–29. https://doi.org/10.1037/h0071663

Wason, P. C. (1960). On the failure to eliminate hypotheses in a conceptual task. *Quarterly Journal of Experimental Psychology, 12*(3), 129–140. https://doi.org/10.1080/17470216008416717

Zewe, A.. (2025, January 17). *Explained: Generative AI's environmental impact.* https://news.mit.edu/2025/explained-generative-ai-environmental-impact-0117

4 Creating Sustainable Value Through Product Management

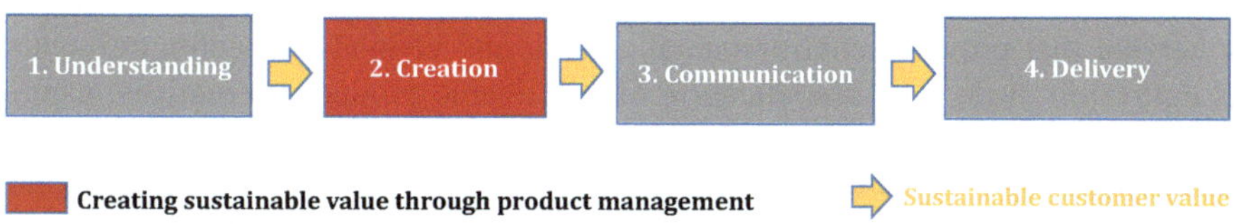

Chapter Introduction
In a world increasingly shaped by environmental realities and evolving social expectations, businesses face an urgent call to rethink how value is conceived, created, and sustained. Product management, once focused primarily on quality or profitability, now sits at the intersection of innovation and responsibility. The challenge is no longer just to deliver what customers want, but to do so in ways that protect the planet, support communities, and ensure long-term economic viability. This chapter explores practical strategies for designing products that solve real problems, simplify lives, or extend their usefulness through durability, circularity, and resource efficiency. We will also highlight the role of sustainable innovation—whether incremental or radical—and the tools, such as life cycle assessment, that enable decisions grounded in a holistic perspective. By integrating design principles, creative solutions, and an understanding of global sustainability trends, this chapter provides a roadmap for product managers and business leaders. Its goal is to inspire a shift from compliance-driven practices to value-driven strategies that allow organizations to thrive while generating tangible, positive impacts on society and the environment.

D. K. Stoyanov, *Sustainable Marketing*, Springer Texts in Business and Economics,
https://doi.org/10.1007/978-3-032-16498-8_4

4.1 Sustainable Product Options: An Overview of Benefits

Although the concept of a truly sustainable product remains open to debate—since all products require energy, water, and materials for their design, production, and use—in the context of the sustainable marketing process presented in the previous chapter we can generally define a sustainable product as any physical good or service that provides environmental, social, and economic benefits throughout its entire life cycle. The classification of sustainable product alternatives inspired by Weybrecht (2010) and presented below is not intended to be exhaustive but rather to highlight key directions in which companies can improve their portfolios. While some categories may slightly overlap, it is essential to consider the broadest possible perspective to understand the various types of benefits a product should offer to be considered sustainable.

- **Problem-solving products:** products that solve actual problems to businesses, policymakers, society, and/or the environment instead of artificially creating new ones. For example, if we take a look backward, many products have been historically imposed on the consumers though their social and environmental utility is disputable. For instance, De Beers' 1930s and 1940s marketing campaign introduced the slogan "A diamond is forever," reshaping societal norms by making diamonds a symbol of eternal love and essential for engagements. This demand was not inherent but created through advertising, addressing no tangible societal issue. In contrast, LifeStraw is an innovative product designed to provide safe drinking water by filtering out bacteria, parasites, and other contaminants. Initially developed as an emergency response tool for disaster-affected areas and underserved communities, LifeStraw has since become widely used by outdoor enthusiasts and travelers. Its portable and affordable design addresses critical societal challenges by reducing waterborne diseases and eliminating the need for single-use plastic bottles, contributing to both public health and environmental sustainability.
- **Life-simplifying products:** products that significantly simplify consumers' lives instead of making it more complicated. *Google Gesture*, for instance, is a concept designed to translate sign language into spoken or written words in real time, making communication easier for the hearing-impaired and fostering inclusivity. The *Fontus water* bottle offers a practical solution for hydration by generating drinkable water from air using solar power and condensation, especially benefiting outdoor enthusiasts or those in water-scarce areas. Similarly, *Unroll.Me* streamlines email management by consolidating subscription emails into a single daily digest and enabling users to unsubscribe from unwanted emails effortlessly, reducing digital clutter and saving time.
- **Multifunctional products:** these are products that provide holistic solutions by combining multiple features into one, eliminating the need for separate single-purpose products. For example, a versatile vacuum cleaner that works

effectively on carpets, hardwood, and tiled floors replaces the need for specialized vacuum cleaners tailored for each surface. Other examples include all-in-one printers that handle printing, scanning, copying, and faxing, or smartphones that integrate functions like cameras, GPS, and music players, replacing the need for separate devices.

- **Circular products:** products that can be easily repaired, rented, borrowed, or recycled at the end of their life cycle. For instance, *Coca-Cola Japan* gives old bottles "second lives" by introducing 16 functional caps that transform bottles into items like spray bottles or pencil sharpeners, making them reusable for various purposes. Similarly, *Epson's in-office paper recycling system*, launched in 2016 in Japan, enables businesses to recycle used paper on-site, turning it into new sheets without the need for extensive external processing. Such innovations contribute to a circular economy by maximizing resource efficiency and minimizing environmental impact.
- **Self-sufficient products:** these are products that require few or no complementary items, helping to reduce the need for additional purchases and promoting environmental sustainability. For example, single-rinse formulations of laundry detergent eliminate the need for extra rinses, saving water and time. Cold-water laundry detergents reduce energy use since about three-quarters of the energy consumption in washing clothes comes from heating water, thus decreasing greenhouse gas emissions. Another example is printers and copiers designed to print on both sides of paper, reducing paper consumption and waste, contributing to both cost savings and environmental sustainability.
- **Harmless products:** products that do not create dependency or health risks, offering safer and more sustainable alternatives to harmful ingredients or substances. For example, aspartame, an artificial sweetener in soft drinks and candies, casein, a milk protein in dairy products, and nicotine in tobacco, are widely used but often lead to health concerns or addiction. In response, alternatives like stevia or monk fruit are replacing artificial sweeteners, while pea and soy proteins offer plant-based options for those avoiding dairy. Additionally, nicotine-free tobacco and herbal smoking blends are emerging as alternatives for smokers seeking to avoid nicotine dependency. These nonaddictive substitutes not only promote healthier lifestyles but also provide more sustainable and safer choices for consumers.
- **Cost- and resource-saving products**: products that save both money and resources for consumers and companies are often based on new technologies that reduce energy or water consumption, leading to long-term cost savings. For example, solar panels allow consumers and businesses to generate their own electricity, reducing reliance on grid power and lowering energy bills. Solar-powered mobile chargers provide a convenient and sustainable way for consumers to charge their devices without incurring electricity costs. Park meters equipped with smart technology can optimize parking fees, reducing the need for additional staff and increasing revenue for cities. Additionally, transparent solar panels can be integrated into windows, generating power while still allowing natural light, helping buildings save on energy costs for lighting and

heating. These innovations not only lower operational costs but also promote sustainability by reducing environmental impact.

- **Durable products:** products that are designed to last beyond the typical lifespan expected by consumers, ensuring they don't break shortly after the warranty expires. An example of this is the slow fashion industry, which emphasizes quality over quantity, producing clothing that is built to last rather than fast, disposable trends. This approach encourages consumers to buy less, but buy better, with items that stand the test of time. Similarly, the "Buy me once" movement advocates for purchasing high-quality products that are durable, repairable, and made to last, offering consumers a long-term, cost-effective solution by reducing the need for frequent replacements. These products not only save money in the long run but also reduce waste and environmental impact.
- **Eco-friendly products:** products that have minimal impact on the environment throughout their life cycle, including production, use, and disposal. These products are designed to conserve natural resources, reduce pollution, minimize waste, and support sustainability. They may incorporate recyclable materials, renewable resources, or energy-efficient processes to lower their ecological footprint. For example, companies producing biodegradable packaging are reducing plastic waste, helping to tackle pollution. *Carbon-neutral products* can be considered eco-friendly products, as achieving carbon neutrality often involves strategies that reduce environmental impact, such as using renewable energy, sustainable materials, and energy-efficient practices. This means the amount of carbon dioxide (CO_2) emitted during their production, distribution, usage, and disposal is balanced by offsetting an equivalent amount of emissions through activities like renewable energy investments, reforestation, or carbon capture technologies. Companies producing carbon-neutral products often focus on reducing emissions at the source by improving energy efficiency, using renewable materials, and adopting sustainable manufacturing practices. Examples include carbon-neutral cosmetics, electric vehicles powered by renewable energy, and furniture made from sustainably sourced, low-carbon materials. These products align with global efforts to combat climate change by minimizing their environmental impact. However, not all eco-friendly products are carbon-neutral, as the broader category of eco-friendliness also includes efforts to reduce waste, water use, and other environmental impacts beyond just carbon emissions. Carbon-neutral products are a subset of eco-friendly products with a specific focus on mitigating climate change by addressing their carbon footprint.
- **Fair products**: products that meet not only legislative but also ethical requirements and therefore go beyond basic legal compliance to address broader moral considerations. For example, while it is legal to purchase coffee grown in regions with low labor standards, Fair-trade-certified coffee ensures ethical sourcing by guaranteeing fair wages and safe working conditions for farmers. This approach supports sustainability and human rights. Similarly, although the sale of traditional gasoline-powered vehicles is still legal, companies like Tesla produce electric vehicles that exceed emissions standards. These vehicles also

address ethical concerns about climate change by promoting cleaner, more sustainable transportation options. Fair products are also fair competition products or, in other words, products developed and marketed in a way that respects intellectual property rights, avoids patent or copyright infringements, and fosters innovation through ethical and legal practices. These products ensure fair play in the marketplace by encouraging originality and collaboration rather than imitation or exploitation of competitors' work. For example, companies that create new technologies or designs by licensing patents or creating original solutions avoid legal disputes over intellectual property. An example could be open-source software, which allows companies to build upon existing frameworks without violating copyrights, while also contributing back to the community. Similarly, design patents for consumer products, like smartphone cases, often ensure that companies develop their own unique designs, rather than copying competitors' protected innovations. By respecting patents and copyrights, companies not only avoid legal issues but also foster innovation in a competitive and ethical manner.

4.2 Sustainable Innovation and Design: A Life Cycle Assessment

4.2.1 Sustainable Innovations

Inspired by Tidd and Bessant (2009), Belz and Peattie (2012) classified sustainability innovations based on two key criteria: (1) *the required knowledge and skills* for implementation—whether they rely on existing expertise or demand entirely new capabilities—and (2) *the field of application*, distinguishing between innovations applied within an existing industry or those that create entirely new markets. Building on this framework, sustainability innovations can be categorized into the following four types:

- ***Incremental sustainable innovations*** involve continuous improvements to existing products, processes, or services to enhance efficiency and reduce environmental impact without fundamentally altering their function. For example, Vestas, a leading wind turbine manufacturer, has consistently refined its blade designs to increase energy output and durability while maintaining the core wind energy technology. Similarly, Tesla's software updates, which improve battery efficiency and vehicle performance over time, demonstrate how incremental improvements can enhance sustainability without redesigning an entire product.
- ***Niche sustainable innovations*** adapt existing sustainable technologies for use in new business fields, addressing evolving consumer and regulatory demands. A notable example is Zipcar, which leveraged car-sharing technology to promote a more sustainable alternative to private car ownership, reducing emissions and urban congestion. Additionally, Vinted, an online marketplace for second-hand

clothing, has expanded the use of digital platforms to promote circular fashion and minimize textile waste.

- ***Alternative sustainable innovations*** replace conventional, resource-intensive technologies with more eco-friendly alternatives. For instance, NatureWorks, a bioplastics company, produces plant-based plastics that serve as an alternative to petroleum-based materials, reducing reliance on fossil fuels. Similarly, Beyond Meat has introduced plant-based meat substitutes that offer a lower carbon footprint compared to traditional livestock farming, demonstrating how alternative technologies can drive sustainability across industries. The innovation of energy floors that collect and transform kinesthetic energy at railway stations, airports, and nightclubs—such as the piezoelectric flooring at Tokyo Station—also fit into this category of sustainable innovation.
- ***Radical sustainable innovations*** represent groundbreaking, disruptive changes that create entirely new markets and industries aligned with long-term sustainability goals. *Climeworks*, a pioneer in carbon capture and storage (CCS), has developed direct air capture technology that removes CO_2 from the atmosphere, addressing climate change in a revolutionary way. CCS is an emerging technology designed to capture carbon dioxide (CO_2) emissions from various industrial sources, such as fossil fuel power plants, heavy manufacturing facilities, and refineries. Once captured, the CO_2 is transported to designated storage sites, where it is securely deposited to prevent its release into the atmosphere, typically in underground geological formations beneath the Earth's surface. Despite its potential to mitigate climate change by reducing greenhouse gas emissions, CCS remains a relatively recent development, and its long-term environmental impacts are not yet fully understood. Concerns include the possibility of CO_2 leakage into groundwater and surrounding ecosystems, potential disturbances such as induced seismic activity, and the overall effectiveness of long-term containment strategies. Further research is necessary to assess these risks and ensure the safety and sustainability of CCS as a viable climate solution. Likewise, hydrogen-powered aviation, championed by companies like *ZeroAvia*, is exploring hydrogen fuel cells as a game-changing alternative to conventional jet fuels, potentially transforming the future of sustainable air travel. Hydrogen-powered aircraft produce zero CO_2 emissions, relying on hydrogen fuel cells to generate electricity and power propulsion systems. However, the widespread adoption of this technology faces several challenges, including the need for substantial infrastructure investment, advancements in hydrogen storage technology, and improvements in fuel cell efficiency. The production of green hydrogen—hydrogen produced using renewable energy—remains costly and energy-intensive, limiting its immediate scalability. Additionally, the aviation industry must address safety concerns related to hydrogen storage and transport, as well as the need for global regulatory frameworks to support the transition. Despite these obstacles, hydrogen-powered aviation represents a promising radical innovation that could significantly reduce the environmental footprint of air travel and contribute to a more sustainable transportation future.

A unique perspective on sustainable innovation comes from the concept of biomimicry—a problem-solving methodology that takes inspiration from nature's processes, structures, and systems to address human challenges (Benyus, 1997). It can be classified into three main types:

- *Innovations inspired by nature's structures and shapes*: this approach mimics biological forms to enhance efficiency and sustainability. An example is the Eastgate Centre in Zimbabwe, designed based on termite mounds' natural ventilation system. By replicating the self-cooling structure of termite mounds, the building reduces energy consumption by up to 90%, minimizing reliance on air conditioning and lowering carbon emissions.
- *Innovations inspired by biological mechanisms and functions*: this involves replicating natural biochemical or physical processes to create sustainable solutions. For instance, biodegradable bioplastics made from shrimp shells or algae mimic nature's method of decomposing organic material. These materials provide an eco-friendly alternative to petroleum-based plastics, reducing plastic pollution and dependence on fossil fuels.
- *Innovations inspired nature's ecosystems and relationships*: this type of biomimicry models entire ecosystems to develop self-sustaining, regenerative systems. An example is circular economy-inspired industrial symbiosis, such as the Kalundborg Industrial Park in Denmark, where waste from one company becomes a resource for another. By mimicking the efficiency of natural ecosystems, this approach minimizes waste, reduces resource consumption, and promotes a closed-loop system for long-term environmental and economic sustainability.

4.2.2 Sustainable Design and Life Cycle Assessment

Life Cycle Assessment (LCA) is a methodology used to objectively evaluate the sustainability impact of a product, process, or service at different stages of its life cycle: *before consumption* (i.e., sustainable production), *during the consumption* (including purchase and use), and *after consumption* (end-of-life) (see Fig. 4.1).

The scope and depth of these assessments can vary widely, depending on the methodological boundaries defined by the analysts. Some LCAs focus narrowly on specific stages—such as raw material extraction or energy use during the product's operation—while others adopt a more comprehensive cradle-to-grave perspective, encompassing all phases from initial resource acquisition to final disposal or recycling. The level of comprehensiveness often determines the reliability and practical relevance of the findings, especially for identifying trade-offs, environmental hotspots, and opportunities for eco-innovation throughout the product life cycle.

A recent study by Liu et al. (2024), which explores the historical development of LCA, notes that the concept was first introduced by *Coca-Cola* in 1969 during

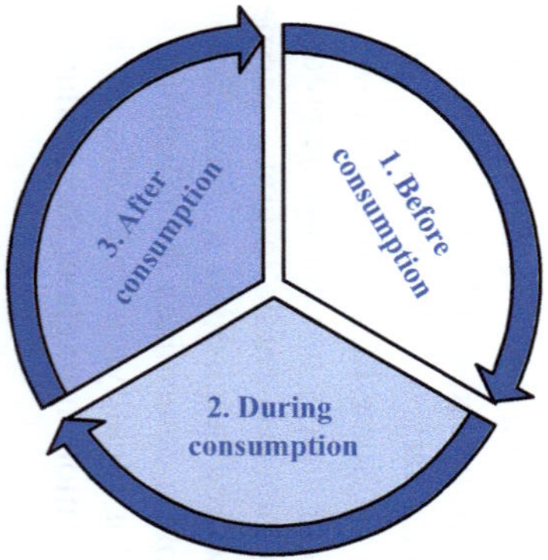

1) Before consumption: at this stage, companies should assess and reduce environmental and social impacts from raw material extraction (e.g., mining, logging and agriculture), material processing (e.g., smelting and refining), component and product manufacturing, packaging, and logistics to distribution centres.

2) During consumption: at this stage, companies should assess and reduce the environmental and social impacts related to the purchase and delivery, product usage (e.g., energy or water consumption), maintenance, repair, consumer experience.

3) After consumption: this stage addresses the product's end-of-life impact. Companies should consider sustainable options like recycling, composting, landfill, incineration (with/without energy recovery), and take-back or refurbishment programs.

Examples of sustainability initiatives :
1. Before consumption: partnering with low-impact suppliers and using certified raw materials; designing products that use fewer materials and are more easily recyclable, safe, and universally accessible; switching to renewable energy in production; measuring and reducing carbon footprint; implementing worker safety programs & supplier audits to ensure labor standards; supporting the development of local communities.
2. During consumption: implementing Product-as-a-Service (PaaS) models to shift from ownership to usage (e.g., leasing clothes); enabling smart usage tracking through IoT devices that optimize product performance and resource efficiency; promoting consumer education and behavioral nudging to encourage responsible use (e.g., water-saving messages); and ensuring transparency in data handling (e.g., GDPR).
3. After consumption: designing products for reuse, recycling, and disassembly (circular economy); adopting Extended Producer Responsibility EPR to take back used products; enabling reverse logistics for refurbishment and recycling; partnering with recyclers and composters for material recovery; and using biodegradable or compostable materials, especially in packaging.

Fig. 4.1 Cradle-to-grave design: actions across product life cycle stages. *Note* The list of initiatives is not exhaustive, and some of them may fit into several categories simultaneously

an evaluation of alternative beverage container options. Despite this early application, LCA remained relatively obscure until it gained broader attention in the late 1980s. The methodology was formally articulated in 1990 by the *Society of Environmental Toxicology and Chemistry* (SETAC) and was later standardized in 1993, when the *International Organization for Standardization* (ISO) integrated it into the ISO 14000 series of environmental management standards. This evolution in LCA methodology has also influenced the development of sustainability-oriented design paradigms, particularly the concepts of *cradle to grave* and *cradle to cradle*. While the former aligns with traditional LCA by focusing on minimizing impacts throughout a product's linear life cycle, the latter advocates for circular systems in which materials are continuously reused, eliminating waste altogether. The *cradle-to-grave* concept was widely disseminated through early LCA practices and formally codified in the ISO 14040 series in the 1990s. In contrast, the *cradle-to-cradle* framework was introduced by McDonough and Braungart (2002) as a more regenerative approach to design. According to their framework, all materials used in industrial and commercial systems should be classified into one of two categories:

(1) ***Technical nutrients*** refer to synthetic or inorganic materials, such as metals and plastics, that are designed for repeated use in manufacturing without degradation in quality. This concept contrasts with traditional *downcycling*, where materials are reused in lower-quality applications—often resulting in eventual waste. For example, a plastic computer casing might be repurposed into a cup, then a phone case, and eventually a park bench, with each cycle diminishing its value and usability.
(2) ***Biological nutrients*** are organic substances that can safely return to the environment through natural decomposition processes. These materials break down into the soil or water without causing ecological harm and serve as nourishment for microorganisms and other life forms. This idea is rooted in the principle of *waste equals food*, where organic waste becomes a resource for decomposers, which in turn regenerate the natural systems that support human life.

Building on the principles of cradle-to-cradle (C2C) design, the Ford Model U—an innovative car prototype developed in 2003 but never commercially launched—demonstrates how sustainable materials and circular design concepts can be applied in automotive engineering. The vehicle is powered by a hydrogen fuel cell, producing almost zero emissions. It incorporates eco-friendly materials such as soy-based foam for the seats, body panels made from a combination of soy and aluminum, sunflower seed-based lubricants, and seat fabrics and dashboards constructed entirely from polyethylene. The design also emphasizes modularity, featuring interchangeable components with minimal variation to simplify assembly, disassembly, and future upgrades. A foldable roof made from corn-based bioplastic further highlights the commitment to renewable, biodegradable materials. Despite these innovations, the Ford Model U was never launched, primarily due to several practical barriers. The high production costs associated with sustainable materials significantly increased the vehicle's overall expense, challenging its economic feasibility. Additionally, the concept required a closed-loop material supply chain—an infrastructure that was difficult to integrate within Ford's existing production system. During the mid-2000s, Ford was also grappling with financial difficulties, prompting the company to prioritize profitable mainstream and hybrid models over experimental sustainability initiatives. Furthermore, regulatory frameworks and industry standards at the time did not require such radical shifts toward fully sustainable design, reducing the commercial urgency for projects like the Model U. This example underscores the challenges faced when attempting to implement cradle-to-cradle principles at scale within established industrial contexts.

However, there have been numerous successful examples of implementing *cradle-to-cradle* principles across various industries, demonstrating the feasibility and benefits of this approach. Companies like *Herman Miller* in furniture manufacturing and *Steelcase* in office equipment have incorporated C2C-certified materials and design strategies to create products that are both sustainable and economically viable. Bioplastics represent another promising application of these principles. Disposable packaging, cutlery, cups, bags, and similar items made entirely from biological materials—such as starch derived from "ugly potatoes"—can be safely returned to the natural environment after use. This significantly reduces the costs associated with waste disposal, including landfill and recycling. These products are designed to be edible, easily dissolved in water, or biodegradable in soil. Additionally, users may have the option to return them for a refund, enabling reuse or proper recycling and helping to close the material loop. Moreover, the establishment of the Cradle-to-Cradle Certified™ Products Program provides a formal certification framework that helps guide and validate the sustainability performance of products based on strict environmental and health criteria. This certification body plays a key role in advancing circular design and fostering widespread adoption of regenerative practices globally. The cradle-to-cradle framework is increasingly recognized as a compelling alternative to traditional sustainability models, such as the 3Rs (reduce, reuse, recycle), by promoting a more holistic and regenerative approach to design (see Fig. 4.2).

4.3 Sustainable Labels

Sustainability labels are marks or certifications on products that provide consumers with information about their environmental, social, or economic impacts throughout the product's lifespan. These labels can be *voluntary*, issued by governments, NGOs, industry groups, or other organizations, or *mandatory*, required by regulatory authorities to ensure compliance with sustainability standards. *Voluntary labels* are applied at the discretion of producers or retailers and are typically verified through third-party certification or self-declaration. Non-compliance carries limited legal consequences, but the labels often signal added value or ethical

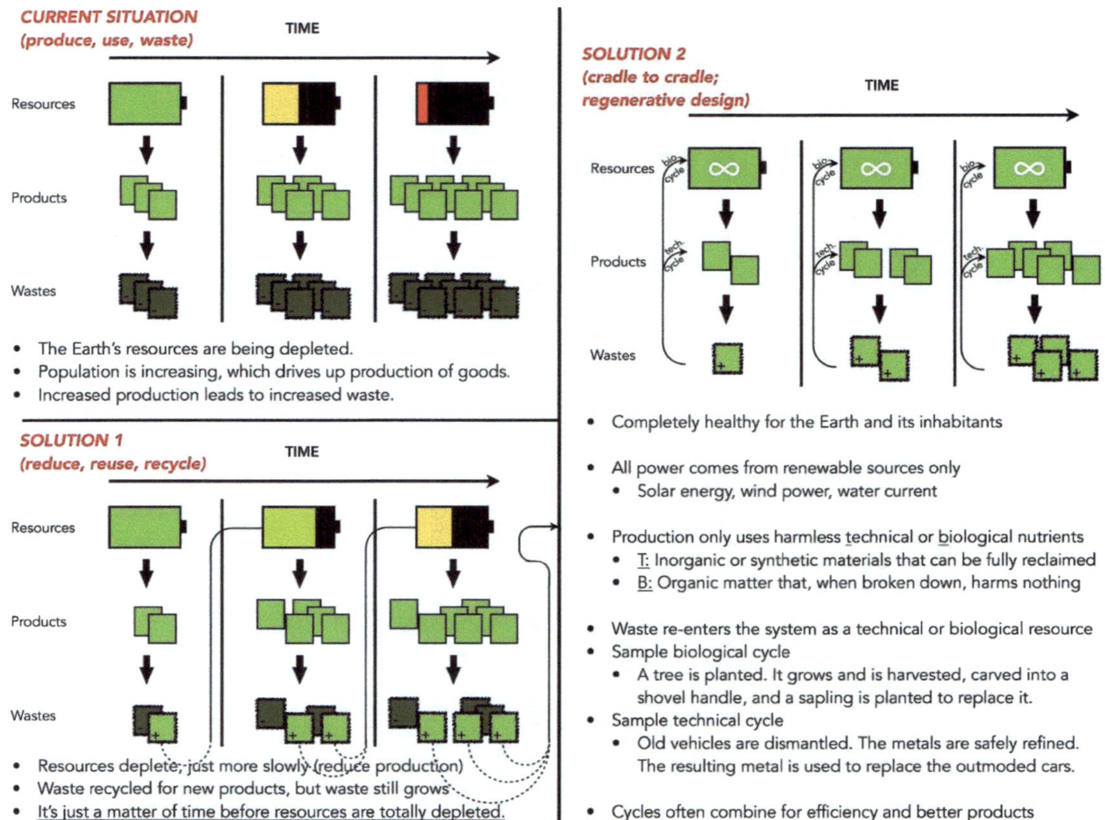

Fig. 4.2 C2C framework as an alternative approach to the 3Rs in sustainable design. *Source* Wikipedia (2017). *Sustainability methods featuring Cradle to Cradle (C2C)*. Retrieved from https://upload.wikimedia.org/wikipedia/commons/b/b1/Sustainability_methods_featuring_C2C.png. Licensed under CC BY-SA 4.0

commitment, helping create niche markets and justify price premiums. Examples include EU Ecolabel, Fairtrade, Rainforest Alliance, and Cradle to Cradle (C2C). *Mandatory labels*, in contrast, are legally required and ensure baseline compliance, providing consumer trust through regulatory oversight but usually without the ability to claim price premiums. Examples of mandatory labels include nutrition and allergen labeling on packaged foods (calories, fat, sugar, salt, allergens), energy efficiency labels such as the EU Energy Label for appliances, and CE marking, which indicates compliance with EU safety, health, and environmental requirements for electronics, machinery, and construction products.

In addition to these sustainability-focused schemes, the European Commission (2025) has developed a system of *geographical indications* (GIs) and *quality schemes* to safeguard the authenticity, quality, and heritage of agricultural products, food, wines, and spirits in the European Union (EU). These schemes protect product names linked to specific regions and traditional know-how, enabling consumers to identify genuine products and allowing producers to differentiate their goods in competitive markets. Registered names are considered intellectual property and play an important role in international trade. All recognized products are listed in official EU registers with details on geographical boundaries, production standards, and specifications. Although EU GIs and quality schemes are not primarily sustainability labels, they can indirectly support sustainability by promoting traditional farming practices, preserving local breeds and biodiversity, protecting cultural knowledge, and strengthening rural economies through price premiums. At the core of this system are *geographical indications*. The *Protected Designation of Origin (PDO)* is the strictest category, requiring all production, processing, and preparation steps to occur in the defined region. For wines, grapes must come exclusively from the region. Examples include Roquefort cheese, which must use milk from a specific sheep breed and mature in the natural caves of Roquefort-sur-Soulzon, France, where the native mold Penicillium roqueforti drives its unique transformation, and Kalamata olive oil from Greece, made entirely from local olive varieties. PDO labeling is mandatory for food and agricultural products and optional for wine. The *Protected Geographical Indication (PGI)* requires that at least one stage of production, processing, or preparation occurs in the region; for wine, 85% of grapes must originate locally. An example is Westfälischer Knochenschinken ham from Germany, cured in Westphalia though pigs may be raised elsewhere. PGI labeling is mandatory for food and agricultural products and optional for wine. The GI for spirit drinks protects spirits whose reputation or characteristics are linked to their region, requiring at least one stage of distillation or preparation locally, while raw materials may come from elsewhere. Irish Whiskey is a common example, and labeling is optional. The *Traditional Speciality Guaranteed (TSG)* scheme protects traditional recipes or production methods without a geographic requirement, ensuring authentic practices are preserved. Gueuze beer from Belgium, made by spontaneous fermentation, illustrates TSG, which is mandatory for all registered products. Two Italian culinary specialties, Mozzarella cheese and Pizza Napoletana, are also officially recognized under the TSG designation. Other EU labels include the *Mountain Product*, highlighting goods from mountain regions with challenging conditions, and the Outermost Regions label, identifying products from remote territories such as Guadeloupe, Réunion, Martinique, the Azores, Madeira, and the Canary Islands. These schemes help farmers market products effectively while informing consumers of origin. Alongside EU-wide schemes, many voluntary national or private certification systems exist, often covering sustainability, fair trade, or animal welfare. The European Commission provides guidelines to ensure transparency and credibility. The system was strengthened by a 2024 reform, introducing a unified legal framework for food, wines, and spirits, simplifying registration, expanding protection to ingredients and online sales, and preventing misuse in digital domains. It also explicitly integrates sustainability, allowing producer groups to include environmental, social, or economic practices in product specifications and voluntarily publish sustainability reports. Producer groups are further empowered to manage, enforce, and promote GIs, increasing their influence within the value chain.

Chapter Conclusion

Creating sustainable value through product management demands a holistic, forward-thinking approach—one that addresses environmental, social, and economic considerations throughout a product's entire life cycle. This chapter highlighted the many facets of sustainable product offerings, ranging from problem-solving and multifunctional designs to the integration of circular and fair products that respect both ecological boundaries and ethical obligations. The discussion of sustainable innovation underscored the spectrum from incremental improvements to radical technologies, championing solutions inspired by nature and guided by life cycle assessments. Through these frameworks, companies can systematically evaluate and reduce their environmental footprints, embracing approaches such as cradle-to-cradle design to promote regeneration and material stewardship. Sustainability labels serve as essential tools for informing consumers and differentiating responsible products in the marketplace. They not only reflect compliance with regulations but also signal deeper ethical commitments and help drive systemic change. Ultimately, sustainable product management is not merely a challenge of design or compliance; it is an ongoing journey of innovation, collaboration, and accountability. By embedding sustainability into every phase of the product life cycle, businesses can unlock new sources of value, remain resilient in changing markets, and contribute meaningfully to a more equitable and regenerative future.

Discussion Questions and Practical Assignments

1. Look on the Internet for the latest energy consumption labels of the EU, South Africa, China, Japan, Australia, and the USA for a specific product category. Make a short table to compare them on the basis of common criteria and discuss from a design perspective what are the similarities and differences between them (e.g., form, shapes, colors, scales, information available). What might be the

Table 4.1 A conceptual framework for comparing energy consumption labels worldwide

Criteria	EU	South Africa	China	Japan	Australia	USA
1						
2						
…						
n						

reasons for the existing differences? Do they reflect a different level of sustainable commitment across nations? (Table 4.1)

Can we have a unique energy consumption label around the world? Why? Why not? If we assume that it is possible, how should it look like? Make a sketch (a simple drawing). What kind of shapes, colors, scale, and informational elements should be present on the label?

2. Research a natural organism or process that exhibits sustainability (e.g., how a lotus leaf repels water, or how bees optimize energy efficiency in flight). Using biomimicry principles, design a product or system that mimics these natural efficiencies, such as creating a self-cleaning surface based on the lotus leaf or a more energy-efficient drone inspired by bee flight patterns. Evaluate your design's environmental impact, its feasibility for real-world use, and its scalability. Develop a marketing strategy to promote your solution and present it in a 5–7 min presentation with supporting visuals.
3. In an era of globalized supply chains and fragmented production systems, does the label "Made in France"—or similar national origin claims—still hold meaningful value for consumers in the twenty-first century? Investigate the legal requirements, advantages, and limitations of obtaining a "Made in [your country]" label, and evaluate whether such labels remain effective in conveying trust, quality, or sustainability. If they do, what specific policies or strategies should governments and institutions implement to ensure their credibility and reduce consumer confusion? If not, what alternative labeling systems—particularly those focused on environmental and social sustainability—could better support consumers in making informed, ethical, and responsible purchasing decisions?
4. In light of growing critiques that the "organic" label no longer fully reflects the environmental or ethical sustainability of food products—particularly in cases such as soilless, chemically supported greenhouse tomatoes—critically evaluate the strengths and limitations of current labels such as "organic," "bio," and "eco-certified" using the tomato industry as a case study. Consider the contrast between heirloom tomatoes, which are open-pollinated, seasonal, and genetically diverse, and hybrid tomatoes, which are mass-produced, seed-dependent, and optimized for shelf life and standardized appearance, often at the expense of taste, nutrition, and biodiversity. Then, propose clear criteria for a next-generation sustainability label that would better inform consumers and promote more responsible production practices, taking into account factors such as seed origin, biodiversity, farming methods, environmental impact, and product seasonality.

5. Research two products within the same category, one with sustainability certifications (such as Fairtrade or C2C) and one without. Compare their marketing strategies, pricing, and consumer perceptions using real-world data or surveys. Summarize your insights in a presentation.
6. Working in teams, design a new product that follows a circular life cycle, ensuring it is repairable, recyclable, or reusable. Develop either a basic prototype or a detailed design brief that explains its functionality and sustainability features. In addition, create a marketing strategy that emphasizes the product's environmental and social value, demonstrating how it appeals to consumers while supporting sustainable consumption and resource efficiency.

References

Belz, F. M., & Peattie, K. (2012). *Sustainability marketing* (2nd ed.). Wiley & Sons.

Benyus, J. M. (1997). *Biomimicry: Innovation inspired by nature*. William Morrow.

European Commission. (2025). *Geographical indications and quality schemes explained*. https://agriculture.ec.europa.eu/farming/geographical-indications-and-quality-schemes/geographical-indications-and-quality-schemes-explained_en#pdo

Liu, M., Zhu, G., & Tian, Y. (2024). The historical evolution and research trends of life cycle assessment. *Green Carbon, 2*(4), 425–437. https://doi.org/10.1016/j.greenca.2024.08.003

McDonough, W., & Braungart, M. (2002). *Cradle to cradle: Remaking the way we make things*. North Point Press.

Tidd, J., & Bessant, J. (2009). *Managing Innovation: Integrating technological, market and organizational change* (4th ed.). John Wiley & Sons.

Weybrecht, S. (2010). *The sustainable MBA The manager's guide to green business*. Wiley.

5 Creating Sustainable Value Through Price Management

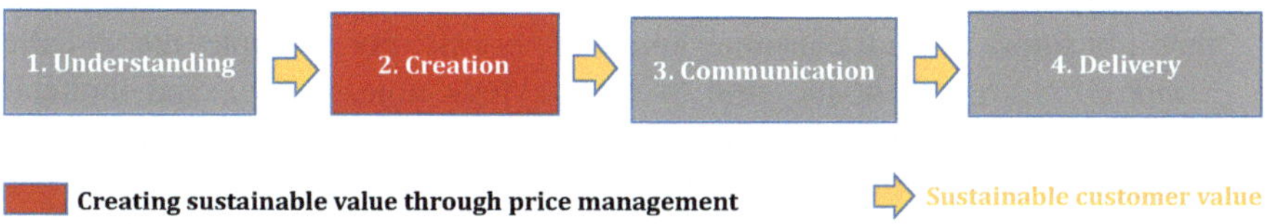

Chapter Introduction

Pricing plays a crucial role in sustainable marketing as it influences both consumer behavior and business profitability. When setting prices for sustainable products, marketers must account for a variety of cost factors that impact consumers, businesses, and society. This chapter explores sustainable price management by examining different cost perspectives, ethical and legal considerations, and emerging sustainable pricing practices. First, we discuss the retrospective and prospective costs that affect consumers' willingness to adopt sustainable products, including sunk costs, opportunity costs, and switching costs. Next, we analyze various unethical and illegal pricing practices that businesses should avoid, such as predatory pricing, price fixing, and bait-and-switch tactics. Understanding these practices is essential for fostering a fair and competitive market. Finally, we introduce a range of sustainable pricing strategies, such as solidarity pricing, social pricing, and pay-as-you-throw schemes, which aim to balance economic, social, and environmental concerns. By the end of this chapter, readers will gain insights into how pricing strategies can contribute to sustainability while ensuring business viability and consumer fairness.

D. K. Stoyanov, *Sustainable Marketing*, Springer Texts in Business and Economics,
https://doi.org/10.1007/978-3-032-16498-8_5

5.1 Sustainable Price Management: A Cost Overview

When deciding on the exact level at which they should set prices of sustainable products, marketers should consider the following types of costs involved in the sustainable production and consumption (See Fig. 5.1):

(1) *Retrospective costs:* these are all the purchase, usage, and post-usage costs that a business or final consumer made previously. For example, the so-called "*sunk costs*" represent the sum of all previous time, energy, and monetary costs related to the consumption of a particular product that may discourage consumers from switching to more or new sustainable product options. They are also known as irreversible costs because they have already been made in the past and cannot be recovered. A classic example are gamblers who do not want to stop playing in a casino because they have already lost a significant amount of their fortune and they believe that it will be the next roll of the dice that will bring them luck. When significant sunk cost has been identified, marketers can offer compensations that are willing to abandon their previous investments (psychological, monetary, and time-related) to switch.
(2) *Prospective costs*: these are the costs expected for the consumers, environment, and society related to the stages before, during, and after product consumption. From an economic standpoint, two types of costs deserve here special attention: (a) opportunity costs— they represent the sum of all potential benefits from an alternative product that a consumer must sacrifice to obtain a product; (b) switching costs— the additional time, energy, and monetary cost that a consumer should make for the sustainability transition or consumption.

One of the major challenges in sustainable pricing is to make consumers aware of all the different costs related to stages before, during, and after consumption, awareness that will help them better understand why the prices of sustainable products are often higher compared to their conventional alternatives and why the purchase of sustainable products should be viewed holistically (as a long-term investment), rather than a transactional purchase. Moreover, different mobile

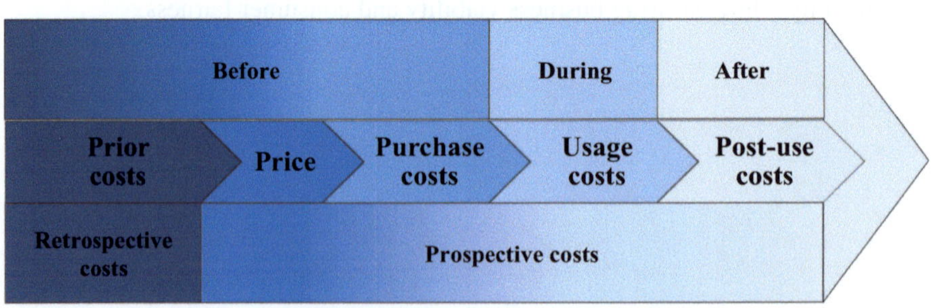

Fig. 5.1 Various types of costs involved in the sustainable consumption and production

applications and web platforms are nowadays dedicated to the estimation of the social, environmental, and monetary impact of businesses, policymakers, and final consumers in an attempt to help them all improve their behavior toward the sustainability direction (e.g., water, energy, plastic, paper, nitrogen, phosphorus, human, land, and biodiversity footprint calculators). For more information on the most frequently used sustainability calculators and the methodology they apply to estimate the different types of footprints (see Čuček et al., 2015). However, when these digital calculators appear on corporate websites, consumers tend to perceive them as less credible and trustworthy, which is why companies often establish partnerships with non-profit organizations. Some of these calculators do not only aim to raise the awareness of the consumers but also to provide them with the opportunity to take actions by compensating their negative impact by making a relevant contribution to NGOs that will in return support various sustainable initiatives worldwide. For example, if you live in Paris, but you want to go on a vacation to Berlin, https://co2.myclimate.org/ shows you that the amount of CO_2 emissions generated from your round-trip flight, in economy class, will be approximately 0.391 t which corresponds to an offset contribution of 10 euros. The German transportation company Flixbus cooperates with a similar-in-scope non-profit organization (*Atmosphair*) to provide its consumers with the possibility to offset their trips (make the carbon footprint of their trips neutral). However, a major problem of these calculators is that they mainly attract consumers that are already aware and committed to sustainable development.

The limitations of such calculators, as well as of all models designed to raise consumer awareness of multiple sustainability-related costs, are consistent with key insights from behavioral economics concerning cognitive constraints and decision-making. Conventional mathematical models developed after the Second World War (e.g., those based on linear, integer, and dynamic programming) generally assumed that economic agents are fully rational decision-makers. Within these models,consumers are expected to carefully process all available market information in order to select the option that optimally satisfies their needs—typically framed in terms of utility maximization for consumers and profit maximization for firms, or the minimization of costs and risks. In reality, however, consumers are not always fully rational in their choices. Even when they strive to behave rationally,they may still be subject to what Herbert Simon (1990) referred to as bounded rationality. This concept reflects the fact that human cognitive abilities are limited and that decision-making is constrained by imperfect information, limited time, and restricted computational capacity. As a result, access to complete or highly detailed information does not necessarily improve decision quality and may, in some cases, impair it—particularly when information complexity exceeds consumers' cognitive resources. Under such conditions, individuals often behave as satisficers, selecting options that meet acceptable or minimum requirements rather than identifying the optimal or best alternative. For example, in a sustainability context, confronting consumers with extensive information on the monetary, environmental, and social costs incurred before, during, and after purchase may overwhelm their cognitive capacities, reducing their ability to effectively evaluate and compare alternatives and resulting in suboptimal decisions. Conversely, when consumers focus exclusively on the monetary cost at the point of purchase (i.e., price), while neglecting other monetary, energetic, environmental, and social costs occurring before and after consumption, their decisions may also be far from optimal from a broader welfare perspective. Similarly, when sustainability labels are overloaded with textual information and pictograms,consumers who must still make a decision may rely on satisficing strategies rather than systematic optimization. Herbert Simon illustrated the decision-making process using the metaphor of a pair of scissors, whose two blades represent cognition (i.e., an individual's limited capacity to process information) and the environment (i.e., the decision context, including information availability and time constraints). Decisions emerge from the interaction between these two elements. The metaphor is particularly apt given that the verb decide originates from the Latin decidere, meaning "to cut off," implying that once a decision is made, alternative options are excluded. Consequently, even seemingly simple decisions can have broader implications for society and the environment [Author's Note]. Consistently, Kahneman (2012) distinguishes between two cognitive systems that shape human thinking and decision-making: System 1, which operates quickly, automatically, and intuitively with little or no effort, and System 2, which is slower, deliberate, analytical, and cognitively demanding. Under conditions of limited time, information overload, or high cognitive demands—common in sustainable consumption contexts—individuals tend to rely more heavily on System 1, helping to explain why real-world consumer behavior often deviates from the predictions of fully rational decision-making models.

5.2 Ethical and Legal Issues in Price Management

Some of the most common unethical and/or illegal pricing practices that sustainable businesses should avoid are presented below:

- **Predatory pricing** — it occurs when a company decides to sell its goods or services below their production costs in an attempt to destabilize or eliminate the competition on the market. Although below-cost pricing is against the antitrust laws established in developed countries, it may happen often and last for a while before public authorities take some measures to stop it, because it is hard to find evidence that an established price reduction deliberately aims at undermining the competition rather than being a part of the normal market competition. For instance, engaged in an intensified price competition on the US market, in 2009 Walmart and Amazon started selling certain hardback best sellers (e.g., those of John Grisham and Stephan King) at prices 50% lower than the suggested list price, because they could afford generating to a certain extent loses while expecting to sell additional products in the long run. Though beneficial for the consumers, this price war significantly weakened the market positions of local booksellers like Barnes & Noble and not surprisingly Amazon and Walmart faced serious allegations of predatory pricing.

- **Dumping** — it is a form of predatory pricing typical for international marketing settings. It occurs when a company exports a product to another country at a price below its production costs in an attempt to undermine the local competition on the foreign market and occupy a dominant position on it. For instance, in an attempt to more easily penetrate the French market, Amazon ignored the anti-dumping regulations established in the country and in August 2000 started selling books with a free delivery option which negatively impacted the financial performances of the local bookstore chains and therefore the US retailer faced severe dumping allegations. Amazon had to pay daily fines of €1000, which, however, did not discourage it to provide free shipping to its online consumers, a predatory pricing practice that lasted till 2014 when the French policymakers officially adopted a law prohibiting free shipping on books.
- **Price fixing:** — it is a secret price agreement (collusion) between two or more parties (i.e., often between companies operating in the same or related industries) that aims at increasing, decreasing, or stabilizing the market price of a particular product or product category for a certain period of time. It is an illegal pricing practice in developed countries as it is in direct contradiction with one of the foundational principles of free market economies—the one stating that the price of a product should be freely determined by the market forces (i.e., the law of supply and demand). However, it is still quite common in oligopolies—markets with imperfect competition dominated by a small number of large market players (e.g., steel production, telecommunications, pharmaceuticals, electric power, and transportation). An interesting paradox is that price fixing is considered illegal when a cartel agreement is made between two or more businesses, but not necessarily when it is intergovernmental (e.g., OPEC). Price fixing can occur both from a seller and a buyer's perspective. For example, on the demand side, some large hospitals can reach an agreement (e.g., a written, verbal, or simply implied one) to purchase only medical supplies from pharmaceutical companies whose prices do not exceed a certain threshold which will put a significant amount of pressure on the pricing decisions of pharmaceutical companies. On the supply side, some of the major manufacturers of pharmaceutical products can agree to fix the prices of their flagship medical supplies to significantly influence hospitals' purchasing decisions.
- **Bid rigging** — it is a special case of price fixing where two or more parties involved in a competitive bidding process decide to either refrain from bidding or to submit quoted prices that are artificially inflated, so that in both cases another participant has higher chances to win. For example, in a bidding process for a construction or software development project, Company A and Company B secretly agree that Company A will submit a higher bid, giving Company B a better chance to win. Additionally, Company C colludes with Company A to inflate their bidding prices. As a result, Company C secures the contract with an artificially low bid, undermining fair competition and potentially leading to financial losses for the entity seeking bids.

- **Price gouging** — it is an abusive pricing practice in which a company sets the prices of its products at a level much higher than the reasonable or fair one (also known as "excessive pricing") but as consumers have fewer options available they can finally still pay the higher price. Sudden price increases can occur in case of supply shortages (e.g., due to natural disasters, global pandemic, or Brexit), when significant gaps between the supply and demand emerge on the market. For example, after the hurricane Sandy hit the USA in 2012, some gas station and car services increased much their prices, gaining vast profits off the desperation of their customers.
- **Price skimming** — an unethical but not necessarily illegal pricing strategy, in which a product is launched by a company at a premium price (e.g., for products positioned as innovations and up-scale products) to enable the company to generate profits quickly enough before its competitors find alternative solutions (substitutes) to lower the market price. As for us, price skimming is justified when significant investments have been made by a company to make a product (e.g., several years in R&D, introduction of revolutionary technology or product feature that satisfies consumers' needs differently compared to other product options available on the market) but not necessarily when the premium price is simply due to significant advertising costs (e.g., luxury brands). When the first iPhone was launched in 2007, it was a radical innovation that completely changed the rules of the game on the mobile phone market. At that time, the starting price for an iPhone of $499 was justified. However, as for us, the iPhone X model, as an incremental innovation launched in 2017, at a starting price of $999 (with features similar and even less performant than other market players) was overpriced.
- **Unjustified price discrimination** — it occurs when identical goods or services are offered by the same provider at different price levels to the same customer segment. For example, airlines, railways, bus operating companies, and hotels often use sophisticated algorithms to track digital consumers and change accordingly the prices based on the amount of time that consumers spend on an online/mobile platform and the number of searches conducted for a particular destination. In 1999 Coca-Cola tested a vending machine that was automatically increasing the prices of soft drink in hot weather but as the concept of dynamic pricing received strong social disapproval the company had to abandon the idea of the discriminatory pricing.
- **Bait-and-switch** — a manipulative up-selling pricing technique that involves a company advertising on a catalogue, TV, or elsewhere one of its products at a highly attractive price (*the bait)* to simply lure consumers into a sales trap at some of its points of sale. When the consumers arrive at the store, the sales force does everything to persuade them to purchase a more expensive alternative (*the switch*) by announcing to them that the product for which they have come has some reliability issues, lacks essential features, or is out of stock. For example, some online retailers are referenced on Google with low prices, as they pretend to sell the cheapest airline tickets just to generate high traffic on their websites (and later sell it to advertisers), but when consumers try to buy their tickets, they

realize that the attractive offers are not available anymore and more expensive offers are recommended to them instead.

- **Captive product pricing** - it is an unethical but not yet illegal pricing practice in which businesses generate higher profitability from selling accessories, consumables, or add-ons in their portfolio, than their flagship products. The idea behind this practice is that a core (main) product serves the role/function of "a hook" (it is provided to consumers either for free or at a very attractive price, sometimes even below the production costs), but the long-term usage or post-usage of the core product is associated with higher costs compared to the initial price paid by the consumer and therefore requires additional regular purchases of more expensive related products. As a result, consumers may feel captivated in their relationship with the company (due to the sunk costs) and refrain from switching to the products of another company. For instance, as a relatively old technology, inkjet printers are nowadays sold in Europe at significantly lower prices compared to laser printers and this is simply because inkjet printers have higher maintenance costs (i.e., for the purchase of ink cartridges) and it is more difficult to recycle, repair, or resell them at the end of their life cycle.
- **Two-part pricing** — a pricing practice often applied in the service industry which involves double charging the consumers—once with an entry fee (called "fixed price") and a second time with a usage fee (called "price per unit"). Depending on the country, it can be considered unethical or illegal when certain taxes are not explicitly communicated to the consumers prior to purchase or when the variable usage fees, without any warning or justification, can largely vary over time. Moreover, similarly to what we discussed on captive pricing, if companies want to build a positive image in the society, they should avoid setting their entry price at a significantly lower level compared to the related usage costs. Some of the companies that often rely on double pricing and can be accused of setting misleading prices are banks, airports, amusement parks, telecommunication, and insurance companies, especially those serving the low-cost segment. For example, low-cost companies like Gold Car have already been accused in France and Spain for proposing car rentals at misleadingly low entry prices (e.g., up to twice cheaper than a regular car rental) when the related usage and post-usage costs of their vehicles (e.g., fuel consumption, daily insurance, deposit guarantee, and repairs) were several times higher compared to a regular car rental.
- **Unfair pricing** — when a company offers significant discounts (e.g., 50%, 70%, or 80% off) on artificially inflated prices. For example, some fast fashion brands launch new collections every month with prices that are initially very high so that after the first week they can be reduced by 50%, after the second week—70%, and before the launch of the new collection—even by 80%. The production costs of those clothes are so low (due to their outsourcing in third-world countries where labor costs are very low) that companies can afford throwing or destroying them without any remorse, simply to make space for the new coming collections.

Pricing practices	Legal	Illegal
Ethical	Solidarity pricing Social pricing	Disobedience of small businesses to excessive taxation compliance
Unethical	Price skimming Bait and switch Captive product pricing Pure-bundling pricing Two-part pricing	Price gouging Predatory pricing Dumping

Fig. 5.2 Pricing practices arising from the opposition between ethics and legislation

- **Pure bundling pricing** — when a firm sells a set of products exclusively as a bundle rather than providing consumers with both possibilities—to purchase each of them individually or altogether (*mixed-bundling pricing*). For example, nowadays consumers cannot buy separately MS Word or MS Excel; they are forced to buy the whole MS Office package which includes many software solutions that not everyone will necessarily need to use or discover.

Figure 5.2 illustrates some of the primary pricing practices arising from the opposition between legislation and ethics in the French market.

5.3 Sustainable Pricing Practices

Ideally, a sustainable pricing practice should meet each of the following criteria:

(1) to be fair for the society
(2) to be fair for the environment
(3) to be fair for the consumer
(4) to be fair for the company

However, there is often only one or few of these criteria that are considered when marketers set the prices for products (See Fig. 5.3).

In the following paragraphs, we present some of the main pricing practices that, as for us, meet to a higher extent the sustainability criteria but still have certain limitations and need further improvement.

5.3.1 Solidarity Pricing

Solidarity pricing, which involves a set of governmental price fixing practices for providing reductions or establishing thresholds such as minimum, maximum, or

Criteria	Fairtrade minimum price *(worldwide)*	Recommended retail price *(Europe)*	Maximum retail price *(India)*
1. Definition	Certification system	Reference system	Gouvernmental policy
2. Level of obligation	Weak	Medium	Strong
3. Product scope	Medium *(agricultural products, textile)*	Medium *(consumer goods)*	Medium *(consumer goods)*
4. Paid by:	Manufacturer/Intermediary	Final consumer	Final consumer
5. Objective	Social (protect suppliers)	Social (protect manufacturer)	Social (protect final consumer)

Fig. 5.3 Sustainable pricing practices from around the world: a comparative perspective

zero-price levels, is often criticized for intervening with the principles of the free market economy, but it has an important role in ensuring wide social access to basic necessities (e.g., working salary, water, electricity, healthcare, and education). For example, *fixed book pricing* policies, already common in developed countries (e.g., France, Germany, and the UK), involve setting prices of books for publishers and bookstores at reasonable levels in an attempt to facilitate the access of the society to information and provide a certain level of protection to books as an important part of the national cultural heritage. Although these pricing practices protect the interests of small bookstores that cannot afford to engage in price wars with major market players (e.g., Amazon and Barnes & Noble), they may not necessarily be beneficial to low-income consumers that cannot afford to purchase high-priced books at the fixed price. *Maximum retail price* (MRP) is the name of the price fixing policy used in India since 1990 to restrict the inflation levels in the country and protect the interests of the final consumers against excessive pricing practices (*price gauging*). It requires manufacturers to indicate on the packaging of the product the highest price at which it can be sold by retailers to the ultimate consumer (MRP) and retailers must not charge consumers over that price. In contrast to the *Recommended retail price* (RRP) which is a reference pricing policy established by manufacturers to simply suggest to retailers the prices at which they should distribute their products, MRP is enforced by law. Some of the major limitations of the MRP are related to the fact that in a way it empowers manufacturers to determine the profit margins of retailers and that retailers can still circumvent the law and make consumers pay a final price in times higher than the MRP by charging them for related services (e.g., a chilling fee applied to the purchase of a soft drink from a vending machine).

5.3.2 Social Pricing

In contrast to solidarity price-fixing practices established by policymakers, social pricing is a form of price fixing initiated by companies. It involves sacrificing the profit margins of some products by proposing them temporarily (*happy hour pricing*) or permanently at low prices (*anti-inflationary pricing*) to make expensive products affordable for consumers with modest income, build good social image, or trigger unplanned purchases. For example, happy hour pricing is a dynamic (variable) pricing policy in which product prices vary according to demand intensity with the most attractive prices being proposed to consumers during periods of low demand to avoid the waste generated when selling perishable, scarce, and seasonal items (e.g., food and drinks in restaurants). It can also be beneficial to businesses that want to generate additional revenues by selling supplementary products via the commercial practices of cross-selling and up-selling. In response to the rising inflation in France, the French retailer Carrefour launched between March and June 2023 an "anti-inflation" basket of 200 essential and nutrition-related products (e.g., detergents, diapers, flour, eggs, milk, and vegetables) with a fixed average price of 2 euros. Some retailers even dare to challenge informal channel leaders in an attempt to achieve social justice. For instance, in 2022, German retailers Edeka and Rewe refused to distribute about 300 well-known international brands (e.g., Twix, M&M, Whiskas, and Coca-Cola), accusing them of benefiting from the inflation by overpricing the consumers. In September 2023, Carrefour placed warning labels on some of its shelves drawing the attention of the consumers to products of Pepsi, Nestle, Lipton, and Unilever that experienced a reduction in quantity instead of a price increase (*shrinkflation*) The slogan on the labels was the following one: "This product has seen its volume or weight fall and the effective price from the supplier rise" (*Ce produit a vu son contenant baisser et son prix augmenter*).

5.3.3 Carbon Offset Pricing

Carbon offset pricing is an environmental pricing practice as it refers to situations in which a company enables its customers to compensate for the main negative environmental impact of their consumption by paying a symbolic tax based on the CO_2 emissions associated with their purchase. Though methane (CH_4), nitrous oxide (N_2O), and fluorinated gases also contribute to the pollution, CO_2 emissions represent about 80% of the global greenhouse gases. *Carbon offset pricing places the environmental* responsibility directly in the hands of the consumers as they are double charged: once for the carbon-related cost of a product and a second time—for the cost of the offset investment. For example, *Atmosfair* is a non-profit organization that provides German citizens with the opportunity to compensate for the greenhouse gas emissions caused by their flights by making donation that aim to contribute to the greener planet www.atmosfair.de. The bus transportation company Flixbus offers consumers to pay a higher price for a trip to compensate for the CO_2 emissions. It varies between 1–3% of the ticket price, so often only a few cents.

5.3.4 Pay-as-You-Throw

Pay-as-you-throw or *PAYT pricing* is a form of *progressive linear pricing (quota-based pricing)* in which policymakers establish quotas for consumers and businesses (e.g., for water and electricity consumption or waste generation) and the more resources they destroy the higher price they pay. The objective is to stimulate the sustainable consumption and production and avoid any potential waste of resources. For example, since 1995, every household and business in South Korea is required to pay a volume-based waste fee rather than a standardized fee at a flat rate. Smart bins are installed all around the country to weigh the amount of waste generated by consumers, which resulted in significant improvements of the recycling rates in the country. In attempt to reduce the excessive water consumption in drought regions deprived from drinking the municipalities countries such as France and the USA have implemented locally progressive linear water pricing policies.

Congestion pricing is also a form of progressive linear/variable pricing that targets excessive users of public goods and services which may be subject to congestion (e.g., bus, airline, electricity, metros, railways, parking, and telephones) by charging them additionally during the periods of peak consumption in an attempt to reduce pollution and traffic congestion. It was firstly implemented in Singapore in 1975, and later adopted by many metropolitan cities in Europe (e.g., London, Stockholm, Milan, and Gothenburg).

5.3.5 Pay-as-You-Use

Pay-as-you-use or *PAYU pricing* is a practice according to which consumers do not have to physically own a product but use it only when they need it (e.g., a self-service laundry, P2P rentals). It opposes the conventional *pay-per purchase* model where consumers are expected to pay the full price of a product and use it whenever and wherever they want. For example, for a construction worker purchasing an electric drill makes sense because the machine will be used frequently (probably on a daily basis) while for individuals that want to renovate their apartment once per 5 years, that kind of machine will be used sporadically (only for making a few holes). However, pay-per-use pricing should often be applied when there is a significant (attractive from a consumer standpoint) gap between full price and the one paid per usage. For example, nowadays, many German hard discounters (e.g., Lidle and Aldi) sell the basic version of an electric drilling machine at a price of 20–30 € when the minimum rental for the same type of machine is 10–15 €. Therefore, *pay-per-usemakes* more sense in the case when essential monetary savings are provided to the consumers (e.g., car purchase vs a traditional car rental vs P2P car rental).

5.3.6 Pay What You Want

Pay what you want or *PWYW pricing* is a practice that involves a marketer charging consumers either a minimum (symbolic) price or absolutely nothing and expecting them to pay the price that they believe the product deserves according to its value. As a part of the value co-creating process that involves higher level of customer engagement in the relationship with the company, this form of co-pricing was already implemented for a variety of entertainment services (e.g., museums, tour guides, software, music, video games, and restaurants). It can be considered sustainable from a social perspective as it aims at protecting the interests of final consumers with modest income. However, it is not applicable for physical goods or products with high marginal costs and in societies where consumers' purchasing decisions are primarily triggered by selfish motives or where they have an extremely low willingness to pay. If PWYW is applied post-factum, it has the potential to significantly reduce the perceived risks of disappointment and cognitive dissonance after the purchase. PWYW can be beneficial for businesses as well. If an element of charity is added to the marketing offer, consumer may be more inclined to make a bigger contribution (donation). Moreover, businesses can benefit from better publicity and higher brand interest, which can eventually improve their positioning strategy and boost their overall sales more than if they simply charge consumers a regular price. For example, in 2007 Radiohead used for the launch of their new album (*In Rainbows*) the strategy of PWYW, where the fans of the band were given the possibility to pay the price they wished for downloading the album. During the first three months when the album was exclusively sold online, 62% of the fans downloaded it for free and 38% paid for it on average $6.00 (Elberse & Bergsman, 2008). Moreover, according to the main vocalist of the Radiohead, as the online sales revenues did not have to be shared with record labels and resellers, they were higher than those of all other albums of the band altogether.

Chapter Conclusion
Sustainable pricing is a dynamic and complex domain that requires balancing economic objectives with ethical, social, and environmental considerations. As explored in this chapter, pricing decisions are influenced by multiple cost dimensions, including retrospective and prospective costs, which impact consumer behavior and market dynamics. Businesses must carefully navigate these costs while fostering a fair and transparent pricing environment. Unethical and illegal pricing practices pose significant risks to both businesses and consumers. Practices such as predatory pricing, price fixing, and price gouging not only distort market competition but also erode consumer trust. By adhering to ethical pricing principles, businesses can enhance their long-term reputation and contribute to a more sustainable marketplace. The evolution of sustainable pricing strategies, including solidarity pricing, carbon offset pricing, and pay-as-you-use models, reflects a growing commitment to sustainability across industries. These approaches demonstrate that pricing can be leveraged as a tool to encourage responsible consumption and production while addressing social

equity and environmental concerns. In conclusion, sustainable pricing is not just about setting the right price; it is about ensuring that pricing strategies align with broader sustainability goals. By embracing fair and transparent pricing models, businesses can create value for consumers, protect the environment, and drive positive societal change.

Discussion Questions and Practical Assignments

1. Use the template provided below to compare at least two of the pricing practices discussed in this chapter and therefore demonstrate their deviations from sustainability. Based on their limitations, make suggestions for improvement. Are you aware of other pricing practices that can be considered to a certain extent sustainable? (see Table 5.1).
2. During the second wave of COVID-19 in Germany, a Mexican restaurant based in Berlin (*Oh La Queca*) has launched the same lunch menu at three distinct price levels, depending on consumers willing to pay (i.e., a solidarity price of 8.25 €, standard price of 10.38 € and support price of 12.50 €). This experiment demonstrated that the majority of the consumers paid the support price. Do you think that this pricing practice can be implemented by fast-food retail chains like McDonalds, KFC, and Burger King for some of their products/at some of

Table 5.1 A conceptual framework for comparing sustainable pricing practices

Criteria	Sustainable pricing practice A	Sustainable pricing practice B
1. Definition	*What is it about?*	*What is it about?*
2. Scope of application	*For what kind of products, services or industries is it applied*	*For what kind of products, services or industries is it applied*
3. Sustainable benefits	*What are the sustainability benefits?*	*What are the sustainability benefits?*
3.1. Social benefits	*What are the benefits to the society?*	*What are the benefits to the society?*
3.2. Environmental benefits	*What are the benefits to the environment?*	*What are the benefits to the environment?*
3.3. Consumer benefits	*What are the benefits for the consumers?*	*What are the benefits for the consumers?*
3.4. Business benefits	*What are the benefits for the companies?*	*What are the benefits for the consumers?*
4. Context of application	*To how many countries, cities, and/or companies can it be applied?*	*To how many countries, cities, and/or companies can it be applied?*
5. Limitations	*What are the potential negative spillover effects on a local, national, or global level?*	*What are the potential negative spillover effects on a local, national, or global level?*

its points of sales and it is worth communicating it as exemplary sustainability pricing practice? Otherwise, what price adjustments would you suggest fast-food retail chains to gain more credibility in the eyes of the sustainable consumers?

3. How can companies balance profitability and sustainability when setting prices for eco-friendly products? Given that sustainable products often have higher production costs due to ethical sourcing, fair wages, and eco-friendly materials, how can companies price them competitively without sacrificing profits? Select a company known for sustainable pricing (e.g., Patagonia, Unilever, Tesla). Research the company's pricing model and explain how it integrates sustainability with profitability. Compare this model to a non-sustainable competitor in the same industry. Discuss whether premium pricing strategies limit market accessibility and propose alternative solutions. Support your argument with case studies and data.
4. What are the psychological barriers that prevent consumers from accepting sustainable pricing models, and how can businesses address them? Conduct a survey or analyze existing consumer research on perceptions of sustainable product pricing. Identify key psychological barriers such as loss aversion, price sensitivity, and perceived fairness. Suggest marketing and pricing strategies that could reduce consumer reluctance (e.g., framing, discount incentives, behavioral nudges). Provide real-world examples of brands that have successfully addressed these challenges.
5. In what ways can unethical pricing practices undermine sustainability efforts? Choose one of the unethical pricing practices discussed in this chapter. Analyze a real-world case where a company engaged in this practice and its consequences. For example, during crises (such as the COVID-19 pandemic), some businesses increased prices for essential goods like masks, sanitizers, and even e-books. How does such behavior erode trust in sustainability claims? Examine consumer reactions using media reports, social media trends, and sales data. Propose policy recommendations or corporate guidelines to prevent unethical pricing in sustainable markets.
6. Use data to assess the effectiveness of various sustainable pricing strategies. Select a sustainable pricing model (e.g., solidarity pricing, carbon offset pricing, or tiered pricing for waste reduction). Gather empirical data from academic papers, business reports, or case studies. Analyze whether the model has led to measurable behavioral or financial outcomes (e.g., increased sales, lower carbon emissions, higher customer retention). Evaluate its long-term sustainability on a global scale.
7. Assess how cultural and regional differences shape consumer attitudes toward sustainable pricing models. Select two or more countries with different approaches to sustainable pricing (e.g., South Korea's pay-as-you-throw model vs. U.S. cities with free waste disposal. Compare consumer responses, government policies, and market acceptance in each region. Analyze how cultural attitudes toward fairness, responsibility, and collective action influence sustainable pricing success. Suggest strategies for businesses to tailor pricing models to different cultural contexts.
8. Examine the distinctions between legal vs. illegal and ethical vs. unethical practices in business, exploring how these categories often overlap or diverge. Analyze

real-world examples, such as price gouging, tax avoidance, or environmental negligence, to determine whether specific actions are legal, ethical, both, or neither. Discuss the implications of these cases on consumer trust, corporate reputation, and regulatory policies. Evaluate whether mere legal compliance is sufficient for ethical business conduct or if companies have a broader responsibility to uphold ethical principles beyond the law. Support your analysis with relevant theories, case studies, and scholarly sources.

References

Čuček, L., Klemeš, J. J., & Kravanja, Z. (2015). Overview of environmental footprints. In *Assessing and measuring environmental impact and sustainability* (pp. 131–193). Butterworth-Heinemann.

Elberse, A., & Bergsman, J. (2008). *Radiohead: Music at your own price (B)*. (HBS Case 9–508–110). Harvard Business School.

Kahneman, D. (2012). *Thinking, fast and slow*. Penguin.

Simon, H. A. (1990). Invariants of human behavior. *Annual Review of Psychology, 41*(1), 1–19. https://doi.org/10.1146/annurev.ps.41.020190.000245

6 Communicating Sustainable Value Through Marketing Communications

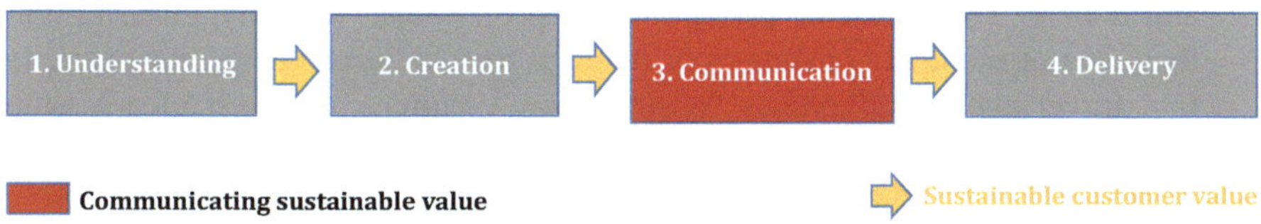

Chapter Introduction

In this chapter, we present the requirements from the 11th edition of the ICC Advertising and Marketing Communications Code, developed by the International Chamber of Commerce (2024), which as for us represent the most detailed and comprehensive guideline on how marketers can communicate sustainable customer value. Specifically, we outline the key principles that should govern responsible marketing practices in advertising, sales promotions, sponsorship, data-driven marketing, environmental marketing, and marketing to children. A more recent edition of the Code can be downloaded from the International Chamber of Commerce website: https://iccwbo.org.

D. K. Stoyanov, *Sustainable Marketing*, Springer Texts in Business and Economics,
https://doi.org/10.1007/978-3-032-16498-8_6

6.1 General Guidelines in Advertising and Marketing Communications

According to the latest version of the ICC Advertising and Marketing Communications Code developed by the International Chamber of Commerce (2024), the following general principles should govern responsible marketing practices:

1. *Fundamental principles*: marketing communications should be legal, ethical, and honest. They must be prepared with professional integrity, ensuring that they do not mislead consumers or damage trust in advertising. For example, if a travel agency advertises "all-inclusive vacations," it must clearly state whether additional fees for excursions, taxes, or tips apply to avoid misleading customers.
2. *Social and environmental responsibility*: advertisements should not encourage discrimination, stereotypes, or harmful behaviors. Marketing should also support sustainable business practices and avoid misleading environmental claims. For example, a cosmetics company claiming its products are "eco-friendly" must ensure its packaging and ingredients meet established environmental standards rather than just using vague, feel-good language.
3. *Respect for decency*: all marketing content should align with social and cultural standards, avoiding offensive or inappropriate material. For example, a perfume brand receives backlash for an overly suggestive ad campaign and revises it to focus on elegance and sophistication rather than sexualized imagery.
4. *Integrity and honesty*: advertising must not exploit consumer trust or take advantage of a lack of knowledge, inexperience, or emotional vulnerability. For example, a home security company should not use fear-based marketing to mislead people into buying unnecessary security services.
5. Accuracy and truthfulness: marketing claims should be accurate and not create false impressions, whether through direct statements, omissions, or exaggerated language. For example, a phone company advertising "unlimited data" must disclose any restrictions, such as speed throttling after a certain usage limit.
6. *Evidence-based claims*: any factual claims in advertisements must be supported by reliable evidence before being published. A skincare brand claiming its product "reduces wrinkles by 50%" must have clinical studies to back up the statement.
7. *Clear identification and transparency*: advertisements must be clearly distinguishable from editorial content, ensuring consumers recognize when they are being marketed to. For example, a social media influencer must disclose when they are being paid to promote a product using hashtags like #ad or #sponsored.
8. *Disclosure of the marketer's identity*: the entity responsible for an advertisement must be identifiable, allowing consumers to contact them if needed. For example, a promotional email should include the company's name, address, and a way to opt out of future messages.

9. *Use of scientific or technical information*: scientific terms, statistics, and research findings should not be manipulated or presented in a misleading way. For example, a toothpaste brand citing "scientifically proven to whiten teeth" must ensure the study referenced is legitimate and not misrepresented.
10. *Appropriate use of "free" offers*: the term "free" should only be used when there are no hidden costs or conditions attached beyond reasonable shipping fees. For example, a streaming service offering a "free trial" must clearly state whether a paid subscription will automatically begin after the trial ends.
11. *Transparency in offers*: all marketing offers should clearly present pricing, conditions, and any relevant restrictions to prevent confusion. For example, a travel agency offering "all-inclusive vacation packages" discloses that flights and taxes are not included.
12. *Fair automatic renewal policies:* subscription services should clearly inform consumers about automatic renewals and provide easy cancelation options. For example, a music streaming service sends a reminder email before the free trial expires, allowing users to cancel before charges apply.
13. *Honest use of guarantees*: if a product is marketed with a warranty or guarantee, the full terms must be clearly communicated. For example, a mattress brand offering a "100-night sleep guarantee" specifies that returns are only valid if the mattress is undamaged.
14. *Fair product comparisons:* comparisons with competitors should be fair, accurate, and based on verifiable facts. For example, a smartphone company comparing battery life with a competitor provides independent testing results in its advertisement.
15. *Respect for brand reputation*: marketing should not unjustly exploit the goodwill or reputation of other businesses. For example, a startup coffee brand cannot claim to be "better than Starbucks" without specific, factual comparisons.
16. *Protection against imitation*: advertising should not copy or mimic another company's branding in a way that confuses consumers. For example, a new fast-food chain is sued for designing its menu and packaging almost identical to McDonald's.
17. *Prohibition of unfair criticism:* marketing should avoid unjustified attacks or ridicule against competitors, industries, or individuals. For example, a soft drink company cannot falsely claim that a rival brand causes health issues without evidence.
18. Testimonials and influencer promotions should be genuine and properly disclosed. A beauty influencer reviewing a skincare product includes #Sponsored and an on-screen disclaimer in their video.
19. *Responsible use of personal identity and property*: marketing materials should not feature individuals or their property without proper consent. For example, a real estate company cannot use a homeowner's picture in ads without permission.

20. *Ethical advertising to minors:* marketing targeted at children and teens must be responsible, appropriate, and not manipulative. A toy company avoids pressure-selling phrases like "convince your parents to buy this now!"
21. *Promotion of safety and health:* advertising should not encourage reckless behavior or downplay health risks. A sportswear company highlights proper usage of equipment instead of promoting "no safety gear needed."
22. *Protection of consumer privacy:* advertisers must respect data protection laws and ensure responsible handling of personal information. For example, a fitness app allows users to opt out of location tracking and targeted ads.
23. *Prevention of unsolicited products and hidden charges:* consumers should not be subjected to unexpected fees or unauthorized product deliveries. For example, a subscription box company cannot send additional products and charge users without prior consent.
24. *Shared responsibility for compliance*: all parties involved in marketing, including advertisers, agencies, and platforms, must uphold ethical standards. A social media platform removes misleading ads about "miracle weight loss" products.
25. *Obligation to correct misleading ads*: if a marketing claim is found to be false or misleading, corrective action must be taken immediately. A fast-food chain corrects an ad that falsely claimed a meal was "low calorie."
26. *Adherence to self-regulatory decisions*: advertisers should comply with industry regulations and respect the rulings of self-regulatory bodies. For example: A car manufacturer updates its marketing after a regulatory body deems its fuel efficiency claims misleading.

6.2 Specific Guidelines on Various Forms of Marketing Communications

In addition to the general principles of responsible advertising and marketing communications, the International Chamber of Commerce (2024) has established specific guidelines to address the unique considerations of various marketing practices, including sales promotions, sponsorship, data-driven marketing, digital marketing, direct marketing, environmental marketing, and marketing to children, as presented in the sections below.

6.2.1 Sales Promotions

Sales promotions must adhere to the following principles to maintain trust and ensure fair competition (International Chamber of Commerce, 2024):

Honesty and fairness: promotions should not be misleading or exploit consumer trust. For example, a retailer advertising a **"50% off sale"** must ensure that prices were not artificially inflated before the discount, preventing deceptive pricing tactics.

Transparency in terms: all conditions, such as eligibility and time limits, must be clearly stated. For instance, a **contest requiring a purchase to enter** should explicitly mention this in the promotional materials rather than hiding it in fine print.

Accurate presentation: promotional offers must not exaggerate benefits or create false expectations. A **"one-month free trial"** that automatically charges users unless canceled should clearly disclose this condition to avoid misleading consumers.

Fair administration: promotions must be executed as promised, with **sufficient stock available** and proper complaint-handling procedures. A **buy-one-get-one-free offer** should not run out of stock too quickly due to poor planning or misleading advertising.

Safety considerations: prizes or promotional items should not pose risks to consumers. For example, a **toy giveaway** should ensure that all items are **age-appropriate and non-hazardous**, preventing potential choking hazards for young children.

Clear consumer communication**:** the actual odds of winning in a sweepstake must be clearly disclosed. A **lottery-style promotion** should not create the impression that everyone will win when, in reality, only a small percentage of participants receive a prize.

Retailer and intermediary responsibilities: retailers and third parties must provide accurate promotional information. For instance, a **supermarket running a coupon promotion** should train employees to apply the discount correctly at checkout, ensuring consumers are not unfairly charged.

Accountability**:** both the promoter and intermediaries share responsibility for ensuring the promotion aligns with legal and ethical standards. For example, a **bank offering "zero-interest loans"** must not include **hidden processing fees,** ensuring full transparency for customers.

6.2.2 Sponsorship

Sponsorship agreements must follow these principles to ensure trust and fair competition:

General principles governing sponsorship: sponsorship agreements must be transparent, ethical, and not misleading, ensuring mutual benefit for both the sponsor and the sponsored entity. For example, Nike sponsoring Serena Williams is an ethical partnership because it aligns with both parties' values in sports performance and women's empowerment, while misleading sponsorships—such as an unhealthy fast-food brand sponsoring a fitness program—can create consumer confusion.

Autonomy and self-determination: sponsorship should not compromise the independence of the sponsored party, allowing them to operate and express opinions freely. For instance, when The Guardian newspaper receives funding from corporate sponsors, it maintains editorial independence, ensuring the sponsor has no influence over its investigative journalism.

Imitation and confusion: sponsorship arrangements should avoid misleading similarities that create confusion between brands, products, or events. A hypothetical example would be a beverage company using similar colors and fonts to Coca-Cola at a FIFA World Cup event to falsely suggest affiliation, even though Coca-Cola is the official sponsor.

"Ambushing" of sponsored properties: companies should not associate themselves with an event without authorization. A real-world example is Puma's marketing campaign during the 2006 FIFA World Cup, where the brand ran football-related ads despite not being an official sponsor, attempting to benefit from the event's visibility without paying sponsorship fees.

Respect for the sponsorship property and the sponsor: both parties must mutually respect each other's brand identity. A real example is Red Bull sponsoring extreme sports events while ensuring the athletes' image is not compromised by excessive commercial branding. Conversely, if a pharmaceutical company sponsored a music festival, it should ensure its promotional content does not negatively affect the festival's atmosphere.

The sponsorship audience: sponsorship should be appropriate and relevant to the target audience. A responsible example is LEGO sponsoring educational programs for children, reinforcing the brand's values in creativity and learning. In contrast, a cigarette company sponsoring a youth-targeted music event would be an unethical and inappropriate sponsorship.

Data capture/data sharing: when sponsorship involves collecting audience data, privacy must be respected. A real example is Formula 1 partnering with technology firms for fan engagement, ensuring that personal data collected for interactive experiences complies with GDPR and other privacy laws. A bad example would be a company collecting customer data through a sponsored event and selling it without consent.

Artistic and historical objects: sponsorship of cultural institutions should respect their integrity. A good example is The British Museum's sponsorship by BP, which has been controversial as it raises ethical concerns about fossil fuel sponsorship of cultural heritage institutions. A better alternative would be a renewable energy company sponsoring an art exhibition on climate change.

Social and environmental sponsorship: sponsorship related to social and environmental causes must be genuine and transparent. A positive example is Patagonia sponsoring environmental conservation initiatives, aligning with its sustainability mission. In contrast, a major polluting oil company sponsoring a "clean energy" campaign without reducing its carbon footprint would be an example of greenwashing.

Charities and humanitarian sponsorship: sponsorship of charitable initiatives should prioritize real impact over self-promotion. A good example is Microsoft sponsoring digital literacy programs in underprivileged schools, providing real educational benefits. A bad example would be a company donating a small amount to a charity but spending more on advertising their "generosity" than on actual donations.

Multiple sponsorship: when an entity has multiple sponsors, each role should be clearly defined. A real example is Formula 1 teams, which have multiple sponsors from different industries (e.g., energy drinks, automotive brands, technology firms), ensuring clear brand representation without conflicts. A potential conflict would be a football club sponsored by two competing betting companies, creating confusion.

Media sponsorship: sponsorship in media should not compromise editorial integrity. For instance, PBS and the BBC accept corporate sponsorships but maintain strict rules ensuring sponsors cannot influence content. In contrast, if a news outlet allowed a major advertiser to dictate editorial content, it would undermine journalistic independence.

Responsibility: Both sponsors and sponsored entities share the responsibility of ensuring compliance with ethical advertising and sponsorship standards. A positive example can be seen in Coca-Cola's sponsorship of the Special Olympics, which promotes inclusivity without making misleading claims. In contrast, an irresponsible case would be an alcohol brand sponsoring a teen music festival, as such practices risk encouraging underage drinking.

6.2.3 Data-Driven Marketing, Direct Marketing, and Digital Marketing

Data-driven marketing, direct marketing, and digital marketing communications should align with the following principles.

Responsibility: marketers must ensure that their data-driven, direct, and digital marketing activities comply with legal and ethical standards, protecting consumers from deceptive, invasive, or misleading practices. A real example is apple's app tracking transparency framework, which requires apps to obtain user consent before tracking their data for targeted advertising. A negative example would be a company using personal data without permission to send unsolicited marketing emails.

Identification and transparency: marketing communications should clearly identify the sender and disclose their commercial intent to avoid misleading consumers. A real example is a social media influencer disclosing paid partnerships using hashtags like #sponsored or #ad. A misleading case would be a brand using fake social media accounts to promote its products without disclosure.

Presentation of the offer: all offers must be presented clearly, without deceptive wording or hidden conditions. For instance, a streaming service advertising a "free trial" should specify if automatic billing starts after the trial period ends. A bad example is a mobile carrier advertising "unlimited data" but failing to mention speed throttling after a certain usage limit.

Transparency on the cost of communication: any costs associated with responding to a marketing communication should be fully disclosed. A real example is a customer support hotline displaying call charges before connecting users. A misleading example would be a contest requiring participants to send premium-rate text messages without informing them of the costs.

Respect for children: marketing communications aimed at children should be responsible, age-appropriate, and not exploit their inexperience or trust. A good example is lego's educational advertisements that focus on creativity without manipulative sales tactics. A bad example is a mobile game targeting children with in-app purchases disguised as game progress.

Respecting the wish not to receive communications: consumers should have the right to opt out of marketing communications easily and without unnecessary obstacles. A real example is the gdpr "unsubscribe" requirement, ensuring that users can opt out of email marketing with a single click. A bad practice would be a company making the opt-out process difficult by requiring users to call a helpline instead of providing a simple online option.

Prices and credit terms: prices, payment conditions, and credit terms should be stated clearly to prevent misleading financial commitments. A responsible example is a furniture store offering "0% financing" with clear terms on the duration and conditions. A deceptive case would be a credit card company promoting "no annual fee" but charging hidden service fees.

Fulfillment of orders: companies must ensure that orders are processed, delivered, and fulfilled as promised. A good example is amazon's guaranteed delivery date system, which refunds shipping costs if an item arrives late. A negative example would be an online retailer advertising "in-stock" products that are actually on backorder, leading to long delays.

Substitution of products: if a product is unavailable, any substitute must be of equal or higher quality and clearly communicated to the consumer before purchase. A responsible example is a grocery delivery service notifying customers and allowing them to approve a substitute item if the original product is out of stock. A deceptive example would be an online retailer automatically shipping a lower-quality alternative without notifying the buyer.

Return of faulty or damaged products: businesses must provide clear and fair return policies for defective or damaged goods. A good example is a clothing brand offering free returns and exchanges for damaged items within 30 days. A bad example would be a company refusing returns on faulty electronics, citing vague warranty exclusions.

Payment and debt collection: payment processes should be transparent, and debt collection practices should be ethical and legal. A good example is a car dealership offering financing with clear interest rates and no hidden fees. A negative example would be a payday loan company using aggressive and misleading tactics to collect debts.

Right of withdrawal: consumers must have the right to cancel purchases within a reasonable timeframe, particularly for online or remote transactions. A responsible example is an online subscription service allowing users to cancel their membership within 14 days without penalties. A bad example would be a gym locking members into long-term contracts with no cancelation options.

After-sales service: companies must provide adequate support and service after a purchase to assist customers with issues or maintenance. A good example is an electronics brand offering free technical support for one year. A poor example would be a home appliance company making it difficult for customers to access repairs or replacement parts.

Telemarketing: telemarketing practices should be ethical, non-intrusive, and respect consumer preferences. A good example is a financial institution only calling pre-qualified customers who have expressed interest in their services. A negative example would be a telemarketing company repeatedly calling individuals who have requested not to be contacted.

Respect for public groups and review sites: companies should not manipulate public opinion by posting fake reviews or suppressing legitimate negative feedback. A responsible example is an e-commerce platform ensuring only verified buyers can leave product reviews. An unethical example would be a hotel paying for positive fake reviews while deleting genuine complaints.

Respecting consumer use of digital interactive media: marketers should respect user privacy and avoid intrusive digital advertising tactics. A good example is a website allowing users to customize cookie settings to control data tracking. A poor example would be a social media app that forces users to watch unskippable ads based on their browsing habits without consent.

Interest-based advertising: targeted advertising should be transparent, with clear opt-out options and ethical data collection practices. A good example is a fashion retailer showing personalized product recommendations based on a user's previous purchases, with an option to disable tracking. A misleading example would be an ad network collecting and selling user data without informing them or providing opt-out options.

6.2.4 Environmental Marketing Communications

Environmental claims in marketing should be truthful, verifiable, and not misleading, ensuring that consumers can make informed choices. The following principles guide responsible environmental marketing communications:

Substantiation: environmental claims must be supported by credible evidence, such as third-party certifications or independent scientific studies. Companies should provide clear proof that their products or practices have a measurable environmental benefit. For example, a detergent brand claiming to be "100% biodegradable" should have lab tests confirming that it fully decomposes without harming the environment. In contrast, a fashion retailer promoting a "sustainable collection" without detailing material sourcing or production practices would be misleading.

Honest and truthful presentation: marketing should not exaggerate or omit key details that could mislead consumers about a product's sustainability. Companies must ensure claims are specific and verifiable. For instance, a bottled water brand promoting "eco-friendly packaging" should specify if it uses recycled materials and disclose its recyclability percentage. A misleading case would be a fast-food chain highlighting its switch to paper straws while continuing to use excessive plastic packaging.

Scientific research: claims based on scientific studies should be accurate, relevant, and not selectively used to create a false impression. If research is cited, it should be peer-reviewed and current. A cosmetic brand promoting "reef-safe sunscreen" should reference independent studies proving its ingredients do not harm marine ecosystems. In contrast, a car manufacturer advertising "zero emissions" while ignoring the carbon footprint of battery production and electricity generation is an example of misleading use of science.

Comparisons: comparative environmental claims should be fair, fact-based, and backed by verifiable data. Companies must avoid vague statements like "better for the environment" without specifying a benchmark. A laundry detergent stating "50% less water usage than the leading brand" should provide independent testing results. Conversely, a clothing company advertising "greener choice" t-shirts without clarifying which aspect is greener—materials, production, or packaging—creates confusion.

Product life cycle: environmental claims should consider the entire life cycle of a product, from raw material sourcing to disposal. If a product is marketed as sustainable, the company should explain its impact at every stage. A carpet manufacturer promoting "recyclable carpets" should specify recycling facilities where consumers can return them. A misleading example is a tech company promoting "sustainable materials" while producing non-repairable devices that contribute to electronic waste.

Claims regarding components and elements: any claims about natural, organic, or recycled content must be specific and quantifiable. If only part of a product meets sustainability standards, the claim should not imply the entire product is eco-friendly. A packaging company stating "made from 80% recycled plastic" provides clear information. However, a snack brand labeling its wrapper as "eco-friendly" when only the outer layer is biodegradable, while the inner layer is plastic, is misleading.

Certifications, signs, and symbols: companies should use only recognized environmental certifications and avoid misleading symbols that falsely suggest sustainability. Third-party verification is essential to ensure credibility. A paper brand displaying the FSC (Forest Stewardship Council) certification ensures it sources wood from responsibly managed forests. In contrast, a cleaning product featuring a self-created "green leaf" logo to imply certification without verification is deceptive.

Environmental attribute claims: claims about carbon neutrality, energy savings, or reduced environmental impact should clearly explain how the benefits are achieved. A hotel chain advertising "100% carbon–neutral stays" should specify whether it uses renewable energy, carbon offsets, or energy-efficient practices. A misleading example is an airline promoting "eco-friendly flights" without addressing the actual carbon emissions of air travel.

Responsibility: the responsibility for truthful environmental marketing lies with both companies and their marketing partners. Brands, advertisers, agencies, and media platforms must ensure environmental claims comply with regulations and do not mislead consumers. A retail company working with suppliers to verify sustainability claims on product packaging demonstrates accountability. On the other hand, a social media platform allowing false "zero-waste" product ads without requiring verification contributes to greenwashing.

6.2.5 Marketing to Children and Teens

Marketing communications targeting children and teens must be responsible, age-appropriate, and respectful of their vulnerability and inexperience. Children are generally defined as individuals under the age of 12–13, while teens refer to those between 13 and 18. The following principles outline ethical considerations in advertising to younger audiences.

General principles: marketing communications targeting children and teens must be truthful, fair, and appropriate for their age and level of understanding, ensuring that messages do not mislead or deceive young audiences. For example, a toy company promoting a playset should accurately depict its features without exaggerating its benefits, whereas a mobile game ad suggesting that all players will easily win rewards, when in fact significant in-app purchases are necessary, would violate these principles.

Exploitation of credulity of children or inexperience of teens: advertisers must take special care not to exploit the natural trust and limited critical skills of children and teens. A responsible example is a kids' cereal brand that provides clear nutritional information without overstating benefits like increased strength or intelligence, while a problematic case would be a retailer using high-pressure tactics in a "limited-time only" sale designed to push children to pressure their parents into purchases.

Avoidance of harm: marketing aimed at younger audiences should steer clear of promoting dangerous or unhealthy behaviors and avoid content that could lead to negative self-image or risky actions. For instance, a sportswear advertisement that encourages children to be active in a safe and inclusive manner is acceptable, whereas a fast-food commercial that portrays junk food as essential to a child's diet can have harmful implications and should be avoided.

Social values: communications directed at children and teens should foster positive social values and inclusivity, rather than reinforcing stereotypes or unhealthy ideals. A beneficial approach is seen in a diverse and inclusive toy campaign that shows children from different backgrounds playing together, while a negative example would be a fashion advertisement that uses heavily edited images to promote unrealistic beauty standards, thereby potentially damaging self-esteem.

Children's personal data: marketers must refrain from collecting personal information from children without obtaining proper parental consent, ensuring that any data used is handled responsibly. An appropriate practice is seen on a children's educational website that requires parental approval before gathering any personal details, while a harmful practice would involve a gaming app secretly tracking and selling children's location data without consent.

Privacy of children and teens: it is essential to protect the privacy of children and teens in all digital engagements, with robust measures in place to safeguard personal information. For example, a social media platform that restricts data collection and targeted advertising for users under 18 exemplifies best practices, whereas a teen-focused online retailer requiring extensive personal details for access to promotions would compromise user privacy.

Responsibility: all parties involved in marketing to children and teens—advertisers, content creators, and media platforms—are responsible for ensuring that their communications adhere to ethical standards. A good example is a children's television network that rigorously reviews all advertisements for age appropriateness, while a concerning scenario would be a streaming service airing unfiltered ads for gambling apps during programs aimed at teenagers.

Other guidelines: additional guidelines require that emerging marketing tactics, such as influencer promotions and in-app purchases, are presented clearly so that children and teens can recognize when they are being marketed to. For instance, a YouTube influencer who transparently discloses sponsorships in language suitable for young audiences is acting responsibly, whereas a mobile game that embeds hidden microtransactions without clear disclosure can mislead and exploit younger consumers.

Chapter Conclusion

The ICC Advertising and Marketing Communications Code (2024) establishes a comprehensive framework to ensure ethical, transparent, and responsible marketing practices across various domains, including advertising, sales promotions, sponsorship, data-driven marketing, environmental marketing, and marketing to children. By adhering to these principles, businesses can foster consumer trust, promote fair competition, and contribute to a responsible marketing environment that benefits both companies and society. Advertising and sales promotions must be truthful, transparent, and fair, avoiding deceptive tactics or misleading claims that could manipulate consumer decisions. Companies must ensure that promotional campaigns clearly present terms, pricing, and conditions while maintaining ethical engagement with their audiences. The integrity of marketing messages plays a critical role in preserving brand credibility and protecting consumers from false or exaggerated claims. Sponsorship must be executed with integrity, ensuring that brand associations do not mislead consumers or create false impressions. Companies should align sponsorships with their values, avoiding partnerships that contradict their sustainability or corporate responsibility commitments. Ethical sponsorship not only benefits the brand and sponsored entity but also reinforces public trust by maintaining transparency and avoiding manipulative branding tactics. Data-driven and direct marketing require strict adherence to privacy, consent, and consumer autonomy. Marketers must respect data protection laws, provide clear opt-in and opt-out mechanisms, and avoid intrusive advertising tactics. Ethical digital marketing ensures that consumers have control over their personal information, preventing exploitation and building long-term trust in brand-consumer relationships. Environmental marketing communications must be substantiated, clear, and free from greenwashing. Companies must provide verifiable evidence for environmental claims, ensuring that sustainability messaging is not misleading or exaggerated. From product life cycle disclosures to responsible use of eco-labels, brands must commit to genuine environmental responsibility rather than using sustainability claims as mere marketing tools. Marketing to children and teens requires additional ethical safeguards to protect their vulnerability and limited ability to critically evaluate advertising. Advertisers must avoid manipulative tactics, respect privacy rights, and promote positive social values rather than reinforcing harmful stereotypes or unhealthy behaviors. Marketing directed at young audiences should prioritize their well-being over commercial interests. Ultimately, responsible marketing is not just a legal requirement but a fundamental obligation for businesses seeking long-term success and consumer loyalty. The principles outlined in the ICC Code (2024) serve as a roadmap for ethical marketing practices that balance corporate objectives with social responsibility. Companies that embrace these standards will not only enhance their reputations but also contribute to a more transparent, fair, and sustainable marketing ecosystem.

Discussion Questions and Practical Assignments

1. Should advertisers be legally required to prove their claims before launching a campaign, or should the burden of proof fall on regulators and consumer watchdogs? How can companies balance persuasive advertising and ethical responsibility without compromising profitability? Is it ever ethical to use fear-based marketing to sell products related to health, security, or insurance?
2. How can companies avoid greenwashing while promoting sustainable products? Should businesses be held accountable for reinforcing harmful stereotypes in advertising, even if such campaigns are profitable? In what ways do socially responsible marketing efforts sometimes serve as a public relations stunt rather than a genuine commitment to change?
3. Should influencers be held to the same advertising standards as major corporations, given their reach and influence? How should companies handle historically misleading advertising campaigns—through public apologies, financial restitution, or policy changes? How does AI-generated content challenge the ethical boundaries of truthfulness in advertising?
4. Should advertisers have free access to consumer data if users voluntarily provide it to social media platforms? Is it ethical for brands to track consumer behavior online for personalized ads without explicit consent? Where should the line be drawn between convenience and privacy infringement in targeted advertising?
5. Should all forms of advertising aimed at children be banned, considering their vulnerability to persuasion? How can marketers balance engaging storytelling with responsible advertising when targeting younger audiences? To what extent

should parents be responsible for filtering unethical advertising content for their children?

6. Find an example of a company accused of greenwashing, such as H&M's "Conscious Collection" or Volkswagen's emissions scandal, and analyze how its environmental claims misled consumers. Discuss the specific claims made, such as H&M's promotion of "sustainable" clothing despite continued reliance on fast fashion practices or Volkswagen's false advertising of "clean diesel" vehicles while secretly using software to manipulate emissions tests. Explain how these misleading claims created false perceptions of environmental responsibility and led to consequences such as lawsuits, fines, regulatory actions, and damaged consumer trust. Finally, suggest how the company could have communicated its environmental impact more responsibly by providing verifiable data, using third-party certifications, and ensuring transparency in sustainability efforts to avoid deceptive marketing.
7. Choose a popular brand and analyze one of its marketing campaigns using the 26 ICC principles. Identify which principles were upheld and which, if any, were violated. Suggest improvements to make the campaign more ethical and socially responsible.
8. Form teams and debate the following resolution: "It is ethical for companies to use personal consumer data to create highly targeted ads." One side will argue in favor of targeted advertising, the other against it. Prepare supporting evidence, real-world examples, and ethical considerations.
9. Select a misleading environmental advertisement and redesign it to comply with ICC guidelines. Submit a before-and-after comparison explaining: the changes made and how they ensure the advertisement is truthful, transparent, and verifiable in accordance with responsible environmental marketing principles.
10. Track the advertisements you encounter in a 24-h period (on social media, TV, websites, etc.). Identify at least three that use questionable marketing tactics. Write a short reflection on how those ads influenced your perception or behavior. Would a less deceptive approach have been equally effective?
11. Imagine you are a chief marketing officer of a major company. Draft a privacy and data protection policy that ensures transparency while still allowing effective digital marketing. Explain how you would gain consumer trust without sacrificing business growth.
12. Should there be a complete ban on marketing directed at children under a certain age, given their limited ability to distinguish advertising from entertainment? Do companies have a moral obligation to protect children from marketing that promotes unrealistic beauty standards, unhealthy food choices, or materialism, or is it solely the responsibility of parents? With the rise of influencer marketing, should child influencers be legally required to disclose sponsorships in a way that is understandable to their audience?
13. Should fast-food brands, alcohol companies, and gambling platforms be allowed to sponsor events targeted at young audiences, such as esports tournaments or music festivals? Is it ethical for educational institutions to accept sponsorship

from large corporations, especially those in industries with potential negative impacts such as fossil fuels, fast fashion, or soft drinks?

14. How do sponsorship deals shape consumer perceptions of brands? Is a company that sponsors environmental or social causes always genuine, or is it just a strategic move for corporate image management? Can a sponsorship be misleading if a company's core business contradicts the values of the event or cause it supports, such as an oil company sponsoring a climate change summit? Should sports teams and athletes be more selective about their sponsors, or should financial considerations outweigh ethical concerns?

Acknowledgments ChatGPT, an AI language model developed by OpenAI, was used in this chapter to support the summarization of the *ICC Advertising and Marketing Communications Code* (International Chamber of Commerce 2024) and to assist in generating discussion questions. The author has critically reviewed, verified, and adapted the outputs to ensure accuracy, relevance, and alignment with the academic objectives of the work.

Reference

International Chamber of Commerce. (2024, September 16). *ICC advertising and marketing communications code* (11th ed.). https://iccwbo.org/wp-content/uploads/sites/3/2024/09/ICC_2024_MarketingCode_2024.pdf

7 Delivering Sustainable Value Through Customer Relationships

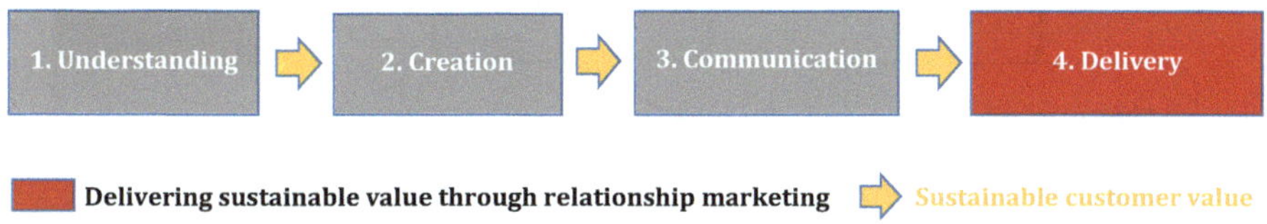

Chapter Introduction

In the rapidly evolving landscape of business and marketing, the concepts of customer satisfaction, experience, loyalty, and trust have undergone significant transformation, especially in the realm of sustainability. As companies shift their focus from traditional product-centric strategies to more holistic approaches that prioritize the well-being of people and the planet, understanding the intricate dynamics between organizations and their stakeholders has become essential. This chapter explores how customer satisfaction has moved beyond simple expectation fulfillment to encompass the entire customer experience, and why acknowledging emotional, cognitive, and behavioral dimensions is crucial for fostering meaningful, long-lasting relationships. The discussion further delves into the dual nature of customer loyalty—attitudinal and behavioral—and the foundational role of trust in building sustainable value. By examining real-world examples and theoretical frameworks, this chapter aims to provide a comprehensive understanding of how managing human relationships underpins successful sustainable marketing and drives organizational growth in an era defined by heightened social and environmental consciousness.

7.1 The Role of Customer Satisfaction, Loyalty, and Trust in Delivering Sustainable Value

The importance of relationship marketing in delivering sustainable value can be broadly explained through the model in Fig. 7.1.

D. K. Stoyanov, *Sustainable Marketing*, Springer Texts in Business and Economics,
https://doi.org/10.1007/978-3-032-16498-8_7

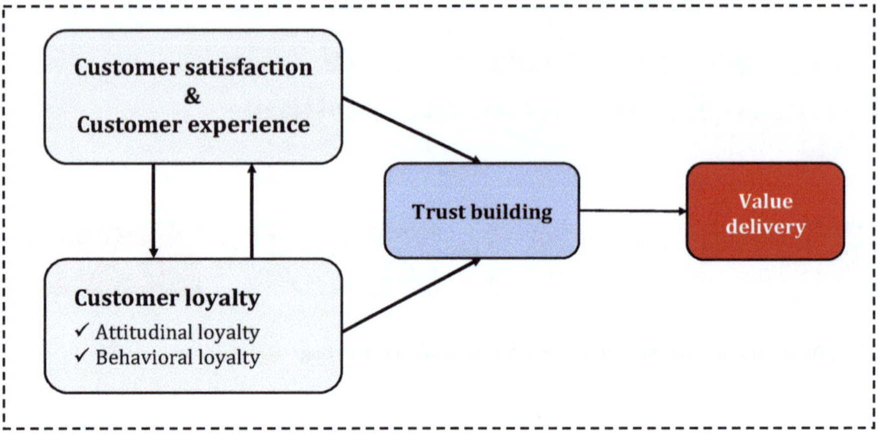

Fig. 7.1 A simplified model of sustainable value delivery in relationship marketing

7.1.1 Customer Satisfaction and Customer Experience

Since the emergence of the marketing concept as a dominant business philosophy in the 1950s, customer satisfaction has been regarded as a primary goal of firms. However, with the rise of the relationship marketing paradigm in the 1980s and the subsequent evolution toward customer experience management, the concept of satisfaction has become less central, giving way to a broader and more holistic focus on the overall customer experience.

7.1.1.1 Customer Satisfaction

Customer satisfaction is the degree to which a product's perceived performance meets or exceeds a buyer's expectations. According to the widely adopted expectation–disconfirmation model, satisfaction results when perceived performance confirms or positively disconfirms prior expectations, whereas dissatisfaction occurs when performance falls short of expectations (Oliver, 1980). Translated into a sustainability context, this implies that customer satisfaction depends on the extent to which a product's environmental and social performance aligns with consumer expectations. For example, customers may expect organic food to be free of harmful chemicals, clothing to be ethically sourced, or electronics brands to offer recycling programs for end-of-life devices. When these expectations are fulfilled or exceeded—such as a company introducing reusable packaging or ensuring transparent fair-trade supply chains—consumer satisfaction with sustainable products is reinforced, encouraging repeated purchases and advocacy for sustainable consumption. Conversely, when expectations are not met, such as when a brand makes exaggerated "green" claims (greenwashing) or fails to deliver measurable environmental benefits, dissatisfaction arises and may discourage consumers from engaging in future sustainable choices (see Fig. 7.2). The model also highlights the

$$Customer\ satisfaction = f(expectations, performance)$$

Product's performance > Consumer's expectations => *Customer delight*

Product's performance = Consumer's expectations => *Customer satisfaction*

Product's performance < Consumer's expectations => *Customer dissatisfaction*

Fig. 7.2 Determinants of customer satisfaction

importance of managing consumer expectations, not just improving product performance. Dissatisfaction may result not only from poor product quality but also from excessively high expectations. For example, a company that heavily promotes its sustainability commitment through intensive marketing across social media, TV, and radio may raise consumer expectations to a very high level. When its products are finally experienced, consumers may feel disappointed, not necessarily because the products are unsustainable, but because the initial expectations created a large gap between perceived performance and anticipated standards. Conversely, a lesser-known brand that communicates its sustainability initiatives sparingly may pleasantly surprise consumers when its actual performance exceeds expectations, thereby enhancing satisfaction.

While the expectation–disconfirmation model has been highly influential, it has several important limitations. First, the model adopts a rational (cognitive) perspective on consumer behavior, assuming that consumers always hold explicit expectations before using a product. In reality, this may only hold for well-known products or incremental innovations, and is less likely for entirely new products without existing alternatives on the market. Second, when consumer expectations are very low or even negative, meeting these expectations does not necessarily result in satisfaction. For example, if a consumer expects that a company will rely on greenwashing to convince them of its sustainability performance and the company does so, the expectation is technically confirmed—but the consumer may still feel dissatisfied. This illustrates that satisfaction is not always a straightforward function of expectations and perceived performance.

7.1.1.2 Customer Experience

Schmitt (1999) notes that traditional marketing focuses on a product's functional features and benefits, treating customers as purely rational decision-makers whose primary goal is to satisfy a need or solve a problem. This approach defines product categories and competition narrowly and relies mainly on analytical, quantitative methods such as regression or conjoint analysis. In contrast, experiential marketing emphasizes creating holistic customer experiences that deliver sensory, emotional, cognitive, behavioral, and relational value. It treats customers as emotional beings influenced by fantasies, feelings, and social contexts, considering the entire consumption process—before, during, and after purchase—as part of a comprehensive experience. Methods in experiential marketing are diverse and exploratory, often employing qualitative or nonverbal techniques, such as eye-tracking, to uncover insights. The ultimate aim is to craft rich, memorable experiences for the customer.

Research by Kumar et al. (2014) and Howell et al. (2012) highlights that consumers increasingly prioritize experiences over material possessions. Spending on experiences—sometimes described as "money spent on doing"—tends to generate more enduring happiness than spending on physical goods ("money spent on having") because the pleasure derived from objects fades over time due to adaptation. Experiences, however, create lasting joy through anticipation, memory, and reflection, while often fostering social connections and enhancing the sense of vitality, leading to greater overall well-being. A sustainability-related example of this principle can be found in eco-tourism or community-based travel experiences. Instead of purchasing luxury items, consumers may spend on guided nature experiences, wildlife conservation activities, or sustainable farm visits. These experiences provide lasting memories, promote social bonding, and strengthen environmental awareness—offering both personal fulfillment and a positive societal impact. In the context of behavioral economics, Kahneman (2010) discusses the so-called "riddle of experience versus memory," noting that people tend to evaluate an experience based more on its peak moments and final outcome than on the overall average of the experience. Applied to sustainability, this implies that even if consumers spend most of their time—say 80%—engaging in a highly satisfying sustainable activity, their overall evaluation will be strongly influenced by the final part of the experience. For example, if the concluding stage of a sustainable service, such as a farm visit or eco-tour, is inconvenient or unpleasant, it could disproportionately reduce their overall satisfaction and likelihood of repeat participation, despite the positive experiences earlier.

Schmitt (1999) identifies five main types of experiences, known as Strategic Experiential Modules (SEMs), each defined by distinct principles, structures, and experience providers:

- *Sense experiences* involve stimulating the five human senses—sight, sound, smell, taste, and touch—to create sensory pleasure, aesthetic appreciation, excitement, and satisfaction.
- *Feel experiences* focus on evoking internal emotions in customers, such as joy, sadness, fear, or anger, with the aim of generating emotional responses.
- *Think experiences* engage the intellect, encouraging creative problem-solving and stimulating cognitive involvement.
- *Act experiences* influence customers' physical behaviors, lifestyles, and social interactions by enriching their daily lives and offering new ways of acting or living.
- *Relate experiences* go beyond the individual, combining aspects of the other modules. They address customers' desires for self-improvement, social connection, and belonging to groups, cultures, or communities.

This experiential classification framework can be extremely valuable when applied to sustainability, as it helps marketers identify gaps and opportunities in sustainability-related experiences. Consider, for instance, a 9D cinema experience where viewers watch an immersive film about global warming and climate change using VR headsets and multi-sensory effects (see Fig. 7.3).

9D cinema experience	Strategic experietial modules								
	Senses					Feel	Think	Act	Relate
	Sight	*Sound*	*Scent*	*Taste*	*Touch*				
Experiential providers	Vivid imagery contrasts thriving ecosystems (lush forests, coral reefs, clean oceans) with environmental degradation (deforested lands, melting glaciers, polluted cities). Snow effects, Rain effects Fog effects	Peaceful natural sounds like birdsong or flowing rivers are juxtaposed with industrial noise, storms, and forest fires.	Fresh scents of flowers, pine, and sea air contrast with smoke from wildfires, polluted air, or oil spills	Bitter taste Salivation	Cool breezes, warmth from sunlight, or ocean vibrations, contrasted with heat from fires or rough textures of degraded land. Leg tickler, Back pocker Butt pocker Water jet Air jet Neck blast Wind effect	Emotions are evoked throughout the experience: empathy for communities affected by climate change, awe at pristine nature, anxiety at environmental threats, and hope through sustainable solutions.	Interactive, mission-based scenarios. Viewers assume the role of protagonists tasked with solving environmental challenges, and analyzing climate data.	Chairs moving in all directions (left, right, front, back, up, down), vibration, effects. Participants grabbing the holding bars tight, laughing, screaming, looking around, closing eyes Moving hands and legs in attempt to defend against the potential threats	Participants develop a sense of connection with each other, much like a crew working together on a shared mission. They also feel linked to the broader goal of sustainable development, fostering a sense of belonging to a community committed to positive environmental action.
Satisfaction	1 2 3 4 **5**	1 2 3 4 **5**	1 2 3 4 **5**	1 **2** 3 4 5	1 2 3 4 **5**	1 2 3 4 **5**	1 2 3 4 **5**	1 2 3 4 **5**	1 2 3 4 **5**

Recommendations for improvement: To engage viewers in a more holistic and pleasing experience, the 9D cinema should, for instance, offer samples of locally sourced, organic, or plant-based snacks or drinks before or after the film. This sensory addition would create a tangible connection between the virtual sustainability message and real-life consumption practices, reinforcing the link between environmental awareness and everyday choices

Fig. 7.3 The application of the strategic experiential modules in a 9D sustainability experience (1). *Source* Own development based on the strategic experiential modules of Schmitt (1999)

7.1.2 Customer Loyalty

Customer loyalty can be broadly divided into two main categories: attitudinal and behavioral, depending on whether it is reflected in a customer's psychological commitment or in their actual purchasing behavior (Day, 1969; Dick & Basu, 1994; Oliver, 1997, 1999). It reflects the degree to which customers not only repeatedly purchase a brand but also develop a favorable attitude and intention to maintain the relationship (see Figs. 7.4 and 7.5). In a sustainability context, this distinction implies that behavioral loyalty is demonstrated when consumers consistently choose sustainable products or engage in sustainable practices, while attitudinal loyalty is reflected in positive perceptions, and advocacy for companies that are genuinely committed to sustainability.

Due to the complexity of the loyalty construct, attitudinal and behavioral loyalty can coexist along the loyalty continuum but with varying degrees of intensity, which makes their assessment and management challenging. Dick and Basu (1994) proposed an integrated model of customer loyalty that enables marketers to distinguish between *no loyalty* (absence of loyalty), *loyalty* (true loyalty), and intermediate forms of loyalty such as *spurious* and *latent loyalty*. These formats help identify gaps in customer commitment and guide the development of targeted strategies accordingly (see Fig. 7.4). Although Dick and Basu's framework was not originally developed for sustainability purposes, it offers valuable insights for enhancing the sustainable value delivery process. For instance, *no loyalty consumers*—those with low attitudinal and behavioral loyalty—are unlikely to respond to new sustainable products, such as reusable packaging, or carbon-neutral services, whose benefits may not be immediately perceived. In markets crowded with conventional or superficially greenwashed alternatives, many consumers prioritize cost and convenience over environmental benefits, making the sustainable value

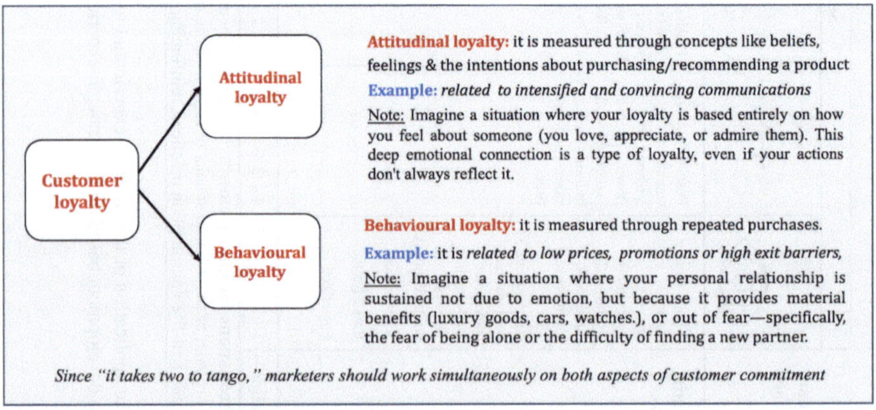

Fig. 7.4 Types of customer loyalty

delivery challenging. Consequently, companies may strategically focus on receptive consumers who are more likely to adopt sustainable products and develop long-term loyalty, maximizing both financial returns and environmental impact. *Truly loyal consumers*—those with high attitudinal and behavioral loyalty—consistently choose a company's sustainable products, resist competitors' offers, and actively share positive word-of-mouth about the brand's environmental benefits. *Latent* and *spurious* loyal consumers, on the other hand, exhibit a mismatch in intensity between attitudinal and behavioral loyalty: one dimension is high while the other is low. Understanding these inconsistencies—arising from habit, subjective norms, or situational factors—is crucial for designing interventions that enhance the delivery of the value proposition. For instance, recent research analyzing 110 studies published in leading marketing journals between 1990 and 2020 provides strong evidence that loyalty programs (LPs) can significantly enhance behavioral loyalty (Belli et al., 2023). While LPs are particularly effective at encouraging repeat purchases and other observable behaviors, shifting attitudinal loyalty remains more difficult. Furthermore, LP effectiveness depends systematically on program design features (e.g., structure, type of rewards, and delivery method) as well as industry characteristics, underscoring the importance of tailoring loyalty initiatives to both the target customer segment and the market context.

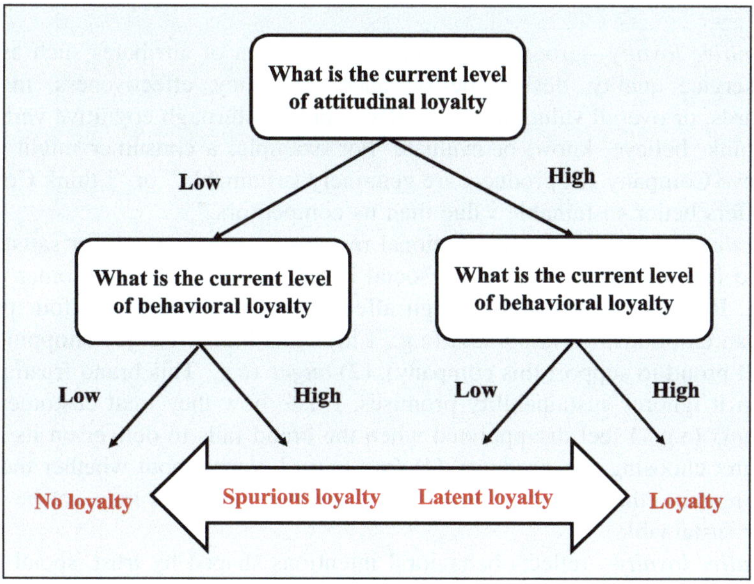

Fig. 7.5 The interconnectedness of attitudinal and behavioral loyalty.
Source Own development based on the classification framework of Dick and Basu (1994)

7.1.2.1 Attitudinal Loyalty

To properly understand the notion of attitudinal loyalty, it is first necessary to define what an attitude is and its core components. In conventional psychology, an attitude is defined as a learned predisposition to respond in a consistently favorable or unfavorable manner to a given object (Allport, 1935; Fishbein & Ajzen, 1975). Attitudes are commonly understood as having three dimensions: (1) *cognition*, (2) *affect*, and (3) *conation* (Eagly & Chaiken, 1993). Cognition refers to the rational component of attitude—what we believe or think about an object. Affect reflects the emotional component—how we feel about it (e.g., like or dislike). Conation represents the intentional or behavioral component—what we plan to do regarding the object, such as buying or not buying. For example: "I am happy when I buy from Company A (*affect*) because it uses renewable energy and fair-trade suppliers (*cognition*), and I intend to keep purchasing and recommending its products (*conation*)." The traditional view suggests that attitudes form in a sequential order: knowledge or beliefs (cognition) come first, followed by emotional evaluation (affect), and then behavioral intention (conation) (Fishbein & Ajzen, 1975). However, alternative perspectives argue that attitudes may form in no fixed order—for instance, one may develop strong emotional attachment to an object without substantial prior knowledge about it (Zajonc, 1980).

Aligned with this perspective, attitudinal loyalty can also be understood as comprising three forms (Oliver, 1997, 1999):

- *Cognitive loyalty*—grounded in rational evaluation of attributes such as product/service quality, design, performance, efficiency, effectiveness, monetary rewards, or overall value. It is typically expressed through cognitive verbs such as think, believe, know, or evaluate. For example, a consumer might say, "I believe Company A's products are genuinely sustainable," or "I think Company A offers better sustainable value than its competitors."
- *Affective loyalty*—based on emotional responses such as customer satisfaction, brand identification, attachment, social influence, or positive customer experience. It is often expressed through affective verbs linked to the four primary human emotions: (1) *happiness* (e.g., I love this brand, I enjoy shopping here, I feel proud to support this company). (2) *anger* (e.g., This brand frustrates me when it ignores sustainability promises, I hate how they treat customers). (3) *sadness* (e.g., I feel disappointed when the brand fails to deliver on its values, I regret choosing this product). (4) *fear* (e.g., I worry about whether the brand will remain ethical in the future, I'm afraid their products may not be safe or truly sustainable).
- *Conative loyalty*—reflects behavioral intentions shaped by trust, social norms, incentives (both monetary and non-monetary), convenience, or commitment to the brand relationship. It is expressed through conative verbs and expressions such as intend, will, plan, commit, or continue. For instance, a consumer might say, "I intend to continue buying sustainable products from Company A," or "I am willing to pay a premium for Company A's sustainable products."

To measure attitudinal loyalty across its different dimensions, marketers typically use classical attitude-measurement scales, such as Likert scales with five, seven, or ten points (Likert, 1932) or semantic differential scales (Osgood et al., 1957). However, in recent years, the Net Promoter Score (NPS) has become one of the most widely used measures of overall, aggregated attitudinal loyalty. NPS is a type of Likert scale, asking customers to rate on an 11-point scale how likely they are to recommend a product or service (Reichheld, 2003). For analysis, responses are typically categorized into three groups—Promoters, Passives, and Detractors—to calculate the overall score (see Figs. 7.6 and 7.7).

NPS is valued for its simplicity and ease of implementation, making it particularly suitable for operational and routine monitoring of customer loyalty. It also enables companies to benchmark their performance against competitors both within the same industry and across different sectors. In addition, NPS provides a practical tool for tracking changes in customer loyalty over time and for guiding strategies to enhance overall customer engagement. Despite its widespread use, NPS has several limitations. First, it relies on customers' stated likelihood to recommend a product or service, but these intentions do not always translate into actual behavior. Second, the scoring system does not fully capture differences within the response range; for example, a score of 0 and a score of 6 are treated similarly, even though they reflect substantially different sentiments. Moreover, the calculation can produce the same NPS for very different distributions of Promoters and Detractors—for instance, 40% Promoters and 30% Detractors yields the same score as 10% Promoters and 0% Detractors (+10)—which may obscure meaningful distinctions in customer loyalty.

Not at all likely	How likely is it that you would recommend this company to a friend or colleagues?*									Extremely likely
0	1	2	3	4	5	6	7	8	9	10
Detractors							***Passives***		***Promoters***	
These are dissatisfied customers who perceive a company's sustainability efforts as inadequate or insincere (e.g., greenwashing). They may undermine the brand's reputation and slow the adoption of sustainable practices by spreading negative word-of-mouth about the company's environmental or social performance							These customers are moderately satisfied with a company's sustainability initiatives but susceptible to switching to competitors that offer better sustainable alternatives, such as products with stronger eco-certifications or transparent ethical sourcing.		These are highly loyal and enthusiastic customers who actively advocate for the company's sustainable practices. By promoting the brand's genuine environmental and social contributions, they generate positive word-of-mouth, strengthen the company's reputation, and encourage wider engagement in sustainable consumption.	
NPS = *Promoters (%) – Detractors (%)*										
Note: *Caution is needed, as some researchers and companies measure likelihood on a 1-to-10 scale rather than the correct 0-to-10 scale. This can bias the results, because starting the scale at 1 implies a minimum 10% likelihood, whereas a respondent may actually intend to indicate 0% likelihood.										

Fig. 7.6 The Net Promoter Score Framework. *Source* Based on the methodology of Reichheld (2003)

Business situation: Let's assume we surveyed a sample of 100 students and asked them: *"How likely is it that you would recommend the quality of sustainable education at your business school to a friend?"* Responses were recorded on a scale from 0 to 10 and presented below:

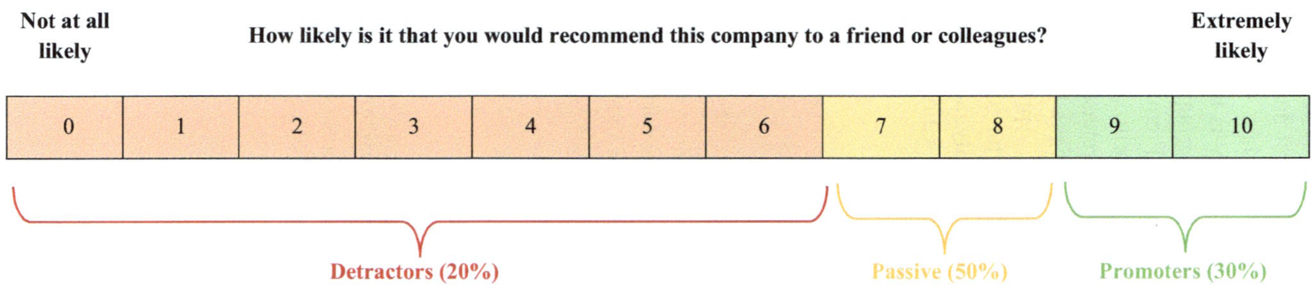

NPS = Promoters (%) – Detractors (%) = 30% – 20 % = 10

Interpretation: The correct interpretation of the school's performance on the NPS requires careful consideration of several factors. For example, if last year's NPS was 5, a score of 10 this year would indicate a significant improvement. Since NPS ranges from -100 to +100, any positive score generally suggests that most students are more loyal than not. However, context is important: the average NPS in the educational sector is approximately 74, meaning that while a positive score is encouraging, it may still fall below industry benchmarks and highlight areas for further improvement. Cultural differences should also be considered, as students from different countries often use rating scales differently For instance, Japanese and Korean respondents tend to be more conservative, rarely giving perfect scores of 9 or 10, whereas higher scores are more common in North America, China, and India, even though these high scores may not necessarily reflect the actual level of commitment. For example, in collectivist cultures such as India and China, individuals may avoid confrontation and prioritize maintaining social harmony, which can influence how they respond to rating scales.

Fig. 7.7 Application of the net promoter score for assessing attitudinal loyalty.
Source Based on the methodology of Reichheld (2003)

7.1.2.2 Behavioral Loyalty

As discussed previously, this is the type of loyalty that is expressed through actual customer behavior rather than attitudes or intentions, and it is therefore not measured through surveys but rather through observed behavioral data obtained from transactional records, CRM systems, or longitudinal panel data. One of the most frequently applied methods for measuring behavioral loyalty is the RFM method which enables companies to categorize their customers into groups such as active, at-risk, or switching customers. RFM stands for Recency, Frequency, and Monetary value, and it allows companies to understand and prioritize their customers according to how recently they purchased, how often they purchase, and how much they spend (see Fig. 7.8). One of the possible applications of the RFM method for measuring the value of sustainable customers is presented in Appendix 1.1.

Other traditional measures of value delivery that reflect behavioral loyalty—and that can be directly applied or slightly adapted for sustainability purposes—include customer retention rate, customer lifetime value, customer equity, and share of wallet. Their applications are presented in Appendix 1.2.

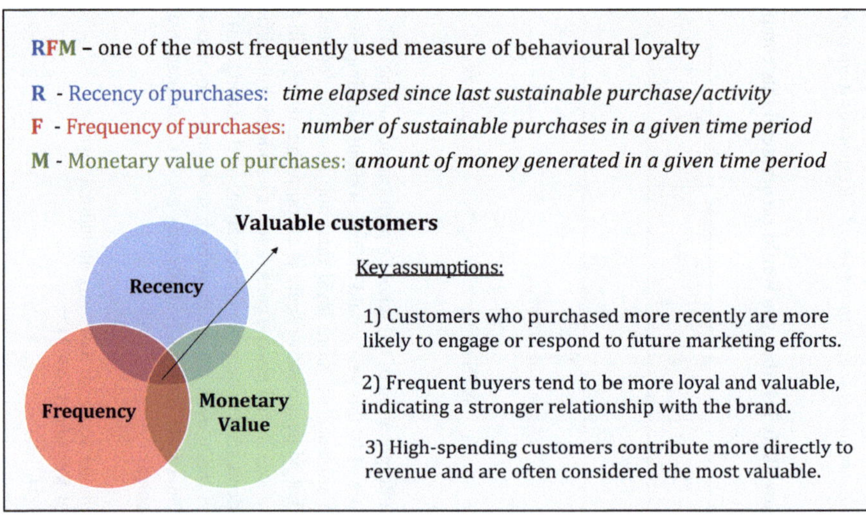

Fig. 7.8 Conceptual framework of the RFM method

7.1.3 Trust Building

According to Maister et al. (2021), trust is a mutual relationship in which one party believes the other is credible, reliable, and caring—qualities that make them willing to be vulnerable or dependent. Trust, they argue, emerges at the intersection of trustworthiness and trusting. In other words, trust is a bi-dimensional concept based on two foundational elements: (1) *Trustworthiness*—the set of qualities that make a business, or individual worthy of trust. (2) *Trusting*—the willingness of one party to be vulnerable and take a risk based on the expectation that the other will act in their best interest. This is an active decision influenced by the perceived trustworthiness of the other party. Even if one party is highly trustworthy, trust cannot exist without the other party's willingness to trust. Conversely, a person may be ready to trust, but if the other lacks trustworthiness, that trust will likely be fragile or short-lived (see Fig. 7.9).

In the context of sustainable marketing, the *trustworthiness equation*—which includes *credibility, reliability, intimacy, and self-orientation*—serves as a valuable framework for building trust with stakeholders who prioritize ethical, social, and environmental responsibility. Trust is cultivated when companies demonstrate genuine sustainability expertise, consistently fulfill their commitments, and form meaningful connections with customers, while placing stakeholder interests above short-term self-interest. However, no matter how high a company scores in credibility, reliability, and intimacy, a high degree of self-orientation can significantly undermine trust and hinder the development of enduring relationships. Striking the right balance among these elements enables marketers to foster long-term, trust-based connections that support sustainable business growth (see Fig. 7.10).

The Volkswagen emissions scandal—widely referred to as "Dieselgate"—offers a cautionary tale of how failing to uphold the elements of the Trustworthiness Equation can severely damage stakeholder trust. Volkswagen promoted its diesel vehicles as environmentally friendly and compliant with emissions regulations. However, it was later revealed that the company had installed software designed to cheat emissions tests, misleading regulators and consumers alike. This revelation shattered Volkswagen's credibility and reliability, as it became clear the company could not be trusted to deliver on its environmental claims. The emotional disconnect that followed, with many customers feeling betrayed, underscored the breakdown of intimacy. Most critically, Volkswagen's actions revealed a high level of self-orientation—the decision to prioritize short-term profits and market share over environmental responsibility and public trust—severely compromising its long-term relationship with stakeholders.

In contrast, Patagonia provides a compelling example of how a company can successfully build trust by aligning with the principles of the Trustworthiness Equation. The outdoor apparel company has established strong credibility through transparent sustainability reporting and clear communication about its environmental impact. It reliably follows through on its commitments, such as using recycled materials and offering repair services to extend product life cycles.

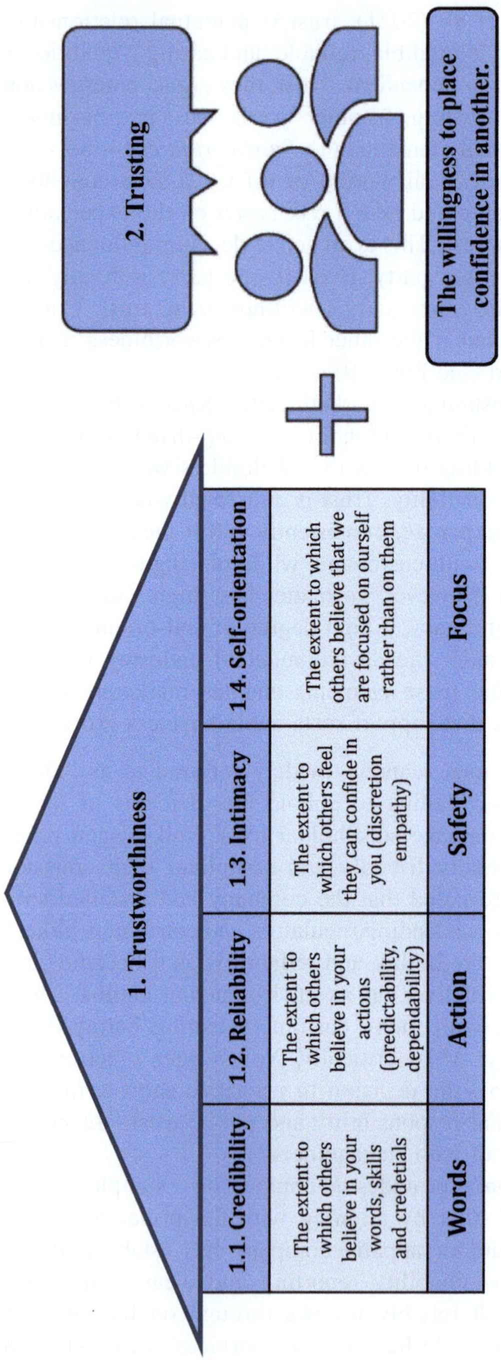

Fig. 7.9 Key elements of trust: trustworthiness and trusting. *Source* Own development based on Maister et al. (2021)

Trustworthiness equation
$$\textit{Trustworthiness} = \frac{\textit{Credibility} + \textit{Reliability} + \textit{Intimacy}}{\textit{Self} - \textit{orientation}}$$
Applications to sustainable marketing :
Credibility - it refers to demonstrating both expertise and honesty regarding a company's environmental and social practices. Brands enhance credibility by communicating transparent, evidence-based claims—such as third-party certifications, verified carbon footprint data, or ethical sourcing policies. For example, publishing audited sustainability reports or aligning with recognized standards (e.g., B Corp, Fair Trade) contributes to perceived credibility. ***Reliability*** – it entails the consistent fulfillment of sustainability-related commitments. Customers expect brands to maintain dependable eco-friendly practices across all touchpoints—such as sustainable product quality, responsible packaging, and fair labor standards. A company that reliably delivers on its green promises across time and contexts reinforces stakeholder trust. ***Intimacy*** – it involves creating an authentic emotional connection with customers by recognizing and addressing their sustainability values and concerns. Brands can cultivate intimacy by engaging in open, honest conversations with their communities, responding to stakeholder input, and involving customers in co-creating sustainable solutions. This two-way interaction deepens relational trust and strengthens brand affinity. ***Self-orientation*** – it refers to the extent to which a company prioritizes its own interests over those of society and the environment. In sustainable marketing, low self-orientation is crucial; it signals a commitment to broader societal and ecological well-being. For instance, a firm that invests in sustainable supply chains or regenerative practices—despite short-term financial trade-offs—demonstrates customer- and planet-centricity, which enhances long-term trust.

Fig. 7.10 The application of the trustworthiness equation in a sustainability context. *Source* Adopted from Maister et al. (2021)

Patagonia also fosters intimacy by aligning its values with those of its customers and involving them in advocacy and environmental activism. Perhaps most notably, it exemplifies low self-orientation by placing social and ecological welfare above profit, as demonstrated by its decision to transfer ownership to a trust that reinvests all profits into environmental initiatives. This long-term, stakeholder-centric approach contrasts sharply with Volkswagen's short-termism, highlighting how balancing credibility, reliability, intimacy, and self-orientation can determine whether a company builds enduring trust—or destroys it.

7.1.4 Value Delivery

The sustainable value delivery is a complex concept in sustainable marketing as apart from its four main dimensions: (1) *value to the society*, (2) *value to the environment*, (3) *value to the company*, and (4) *value to the consumer*, each has tangible and intangible aspects (see Fig. 7.11). From a *social perspective*, sustainable value delivery requires assessing both intangible outcomes—such as those captured through social impact scorecards or stakeholder satisfaction indices—and tangible outcomes, including community investment as a percentage of profit or measurable improvements in local well-being. From an *environmental perspective*, sustainable value can only be achieved by addressing ecological footprints and improving resource efficiency across the entire supply chain network, through tangible indicators such as the carbon footprint per product sold, the percentage of recyclable or biodegradable packaging, and reductions in water or energy intensity, as well as intangible outcomes such as biodiversity. From a *business perspective*, both tangible and intangible benefits are equally critical. Hard performance indicators—such as customer lifetime value, customer equity, or revenue from sustainable products—must be complemented by soft measures that capture perceptions and attitudes, such as the NPS. From *a consumer perspective*, firms can no longer afford the luxury or the moral risk of treating customers merely as transactional targets or of making superficial environmental claims that verge on greenwashing. The era of exploiting customer data, attention, and creativity solely for shareholder gain is rapidly ending. Today's consumers recognize their role as co-creators of value and expect fairness, transparency, and mutual respect in return. Digital platforms and social networks would have little meaning without user-generated content—the very consumers who now demand ethical conduct, authenticity, and sustainability in exchange for their engagement and trust. Together, these four perspectives illustrate that sustainable marketing is not simply about minimizing harm—it is about creating shared, multidimensional value that aligns the interests of business, society, consumers, and the environment in pursuit of long-term well-being.

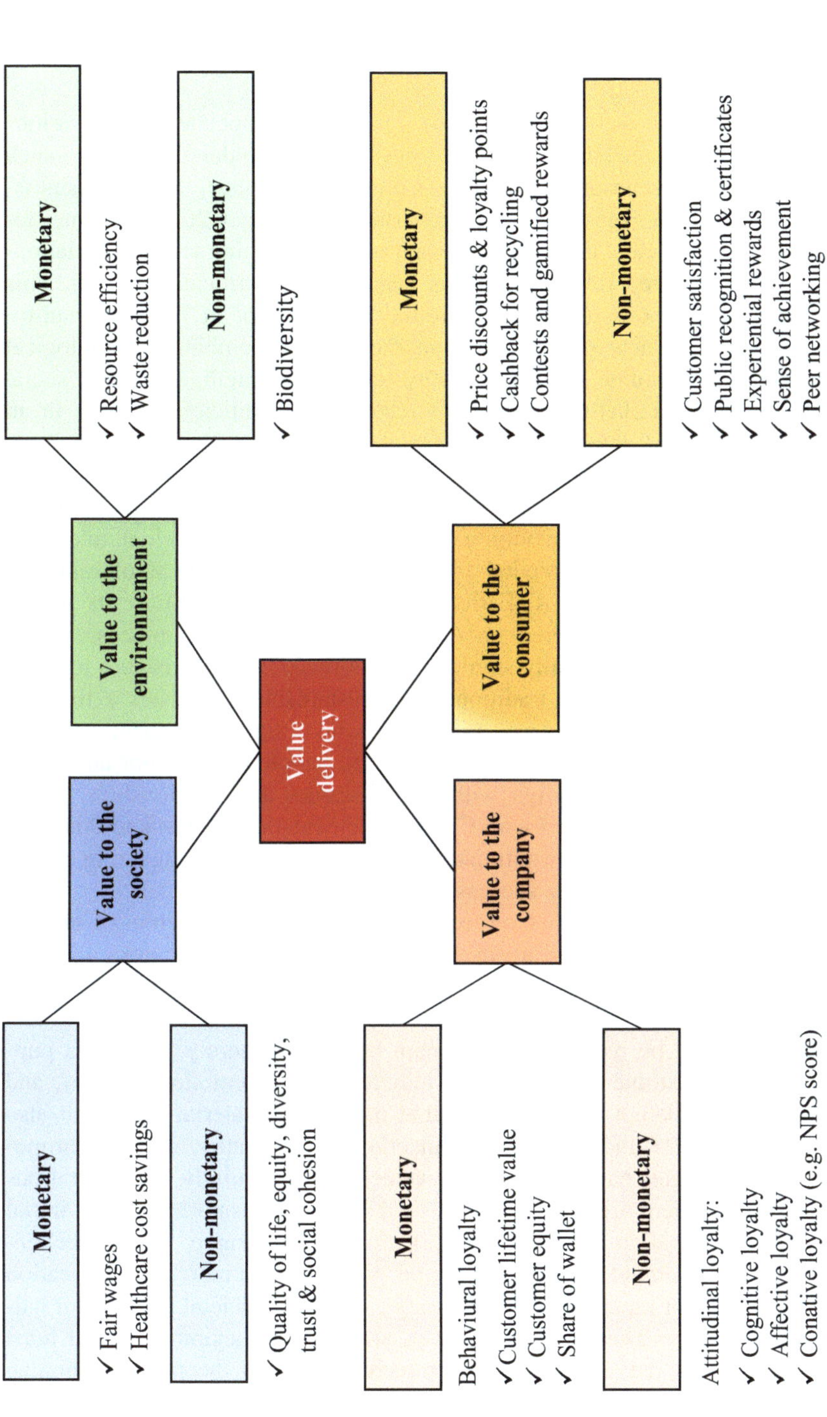

Fig. 7.11 The multidimensional structure of value delivery in sustainable marketing. *Note* The examples of monetary and non-monetary value propositions provided in the bullet points are not exhaustive

7.2 Managing Human Relationships for a Sustainable Future: Concluding Reflections

Marketing as a business philosophy has long been associated with various interaction models, depending on the types of stakeholders involved—such as B2C (business-to-consumer), B2B (business-to-business), B2G (business-to-government), G2G (government-to-government), and C2C (consumer-to-consumer). In recent years, however, the emphasis on managing customer relationships as human-centered relationships has become more critical than ever. This shift inspired Kotler et al. (2021) to introduce the concept of H2H (human-to-human) marketing, a new generation of marketing that combines technological advancement with genuine human empathy, emotional intelligence, and social responsibility. H2H marketing emphasizes restoring authenticity and trust in an increasingly digitalized environment by placing people—not algorithms or transactions—at the center of marketing relationships. Similarly, a recent systematic review highlights Sustainable Customer Relationship Management (*Sustainable CRM*) as the latest stage in the evolution of relationship marketing which integrates social, economic, and environmental factors into the development of mutually beneficial customer relationships (Ferrer-Estévez & Chalmeta, 2023). Its implementation methodology—centered on the alignment of people, processes, and technology—remains conceptually similar to traditional CRM approaches to value delivery. At the heart of both traditional and sustainable CRM lies a timeless truth: people must always come first. It is people who design, implement, and support CRM systems—and customers themselves are people as well. For any CRM initiative to succeed, it must align with human values and relationships. Many organizations that have struggled with CRM implementation overlooked this principle, treating technology as the ultimate solution rather than recognizing it as a tool—important, but not the key success factor.

Moreover, the growing reliance on technologies such as the Internet of Things, artificial intelligence, and other advanced digital systems has paradoxically created a crisis in the delivery of genuine customer satisfaction. Over the decades, marketing practice has, at times, drifted from its foundational principle—that the customer should be regarded as a human being, not merely a pawn, a puppet, or an asset. Customers are complex individuals with emotions, values, and aspirations; they seek not only products that meet their material needs but also relationships that offer understanding, connection, and empathy. While anthropomorphized technologies have enhanced convenience—particularly for Generations Z, Y, and Alpha—there are growing signs of technological saturation. The initial enthusiasm for innovation is giving way to technological fatigue, or even technological disgust. Movements advocating digital detox and the increasing relocation from urban to rural or nature-connected lifestyles signal a societal longing for balance and authenticity. These shifts, however, are only the beginning. What must ultimately change is the way we integrate technology into the fabric of human life—anchoring it in empathy, ethics, and emotional intelligence.

Having spent most of my professional and academic career in marketing, I have observed this transformation firsthand. Before the outbreak of the COVID-19 pandemic in early 2020, students were highly enthusiastic about digital tools and increasingly demanded interactive, technology-enhanced learning experiences, prompting universities and business schools to adopt such solutions on a larger scale. Yet, after the pandemic, the pendulum swung back: students began to value face-to-face interaction, human warmth, and genuine presence over convenience and automation. The experience revealed a timeless truth—technology can enhance, but never replace, human connection. Prior to writing this book, I spent more than fifteen years researching and working in the field of vending retail marketing—a fascinating domain heavily dependent on self-service technologies to deliver efficiency and customer satisfaction. When I began my research, vending retail was in its infancy, and I was among the early scholars exploring its potential (Stoyanov, 2014, 2015, 2021a, 2021b, 2023, 2025). Without any intention of self-praise, I can relate to the reflections of great thinkers such as Albert Einstein, J. Robert Oppenheimer in physics, and Geoffrey Hinton in artificial intelligence—each of whom, post-factum, grappled with the ethical and societal implications of the very innovations they helped create. Although my contributions are far more modest, I arrived at a similar realization: that to truly understand a system, one must also confront its human and moral boundaries. Today, vending technologies are omnipresent—a testament to remarkable technological progress and operational efficiency. Yet this ubiquity comes at a cost: the progressive erosion of human interaction. I have personally experienced this decline in daily life. Whether in a library, a post office, an airport, or any other environment equipped with self-service technologies, it is becoming increasingly difficult to find an actual human being to assist you. Even when such assistance is available, the response is often, "Use the kiosk; I'll show you once, but next time, you can do it yourself." Chatbots now simulate empathy, offering algorithmic approximations of human conversation. When I first began promoting the study of vending retail within both business and academia, I never imagined a future where human contact would be systematically replaced in the name of efficiency and convenience. I fear that these ongoing degradations in human relations—masked by the rhetoric of innovation and progress—will have severe societal consequences in the years to come, challenging not only the essence of marketing but the very definition of what it means to be human.

Some may wonder why emphasizing humanized marketing is essential in a textbook on sustainability. The answer is simple: sustainability begins with humanity. Without recognizing and nurturing the human dimension in our economic, social, and technological systems, sustainable development remains an abstract ideal rather than a lived reality. True sustainability is not just about preserving resources or reducing emissions; it is about cultivating empathy, responsibility, and respect for one another. These are the values that transform marketing from a transactional practice into a transformative force for both human and planetary well-being.

Chapter Conclusion

As we reflect on the principles outlined in this chapter, it becomes evident that sustainable marketing is not merely a matter of promoting environmentally friendly products or ethical business practices—it requires a fundamental reorientation toward human-centered relationships. The evolution from B2C and B2B models to H2H (human-to-human) marketing underscores the necessity of empathy, authenticity, and emotional intelligence in fostering genuine customer engagement. While technological advancements have enhanced our ability to connect and serve customers, a balance must be struck to avoid alienation and technological fatigue. Ultimately, sustainability begins with humanity: only by respecting and nurturing the human dimension can organizations create lasting value for both society and the planet. By prioritizing trust, satisfaction, and loyalty in a holistic manner, marketers can transform their practices into a powerful force for sustainable development, ensuring that economic progress aligns with the greater good of people and the environment.

Discussion Questions and Practical Assignments

1. How has the shift from customer satisfaction to customer experience changed the way marketers deliver sustainable value? Apply Schmitt's (1999) framework of the five strategic experiential modules to design a holistic customer experience that encourages sustainable behavior.
2. Using the Trustworthiness equation developed by Maister et al. (2021), analyze how brands gain or lose trust in sustainability communication by comparing two brands, and identify which component most influences their perceived trustworthiness.
3. What does the shift from B2C/B2B to H2H marketing reveal about the need for empathy and authenticity in sustainability? Draft a one-page *Human-to-Human Sustainability Manifesto* including 3–5 actionable principles for a brand's human-centered communication strategy.
4. In the context of technological saturation, how can firms ensure sustainability messages and experiences retain their human relevance and emotional impact?
5. Provide examples of qualitative (open-ended) and quantitative (closed-ended, 7-point scale) questions that can be used on social media to collect data on attitudinal loyalty—specifically, cognitive, affective, and conative loyalty—toward a sustainability initiative (e.g., a brand's carbon-neutral program, a plastic-free campaign, or a renewable energy project).
6. Use the information below to help "GreenTech B2B Solutions" understand how its long-term industrial customers value their sustainability experience with GreenTech and compare it with other market providers—IBM and Intel (Table 7.1).

Table 7.1 Recommendation likelihood distribution by company

How likely is it to recommend each company in the industry	GreenTech B2B Solutions	IBM	Intel
Extremely likely	20%	40%	5%
	20%	15%	30%
	15%	10%	25%
	10%	10%	0%
	10%	0%	0%
	0%	10%	0%
	0%	5%	0%
	10%	5%	15%
	5%	0%	10%
	5%	5%	10%
No at all likely	5%	0%	5%
Total	100%	100%	100%

References

Allport, G. W. (1935). Attitudes. In C. Murchison (Ed.), *Handbook of social psychology* (pp. 798–844). Clark University Press.

Belli, A., O'Rourke, A.-M., Carrillat, F. A., Pupovac, L., Melnyk, V., & Napolova, E. (2023). 40 years of loyalty programs: How effective are they? Generalizations from a meta-analysis. *Journal of the Academy of Marketing Science, 50*(1), 147–173. https://doi.org/10.1007/s11747-021-00804-z

Day, G. S. (1969). A two-dimensional concept of brand loyalty. *Journal of Advertising Research, 9*(September), 29–35. https://doi.org/10.1080/00218499.1969.12520010.

Dick, A. S., & Basu, K. (1994). Customer loyalty: Toward an integrated conceptual framework. *Journal of the Academy of Marketing Science, 22*(2), 99–113. https://doi.org/10.1177/0092070394222001

Eagly, A. H., & Chaiken, S. (1993). *The psychology of attitudes.* Harcourt Brace Jovanovich.

Ferrer-Estévez, M., & Chalmeta, R. (2023). Sustainable customer relationship management. *Marketing Intelligence & Planning, 41*(2), 244–262. https://doi.org/10.1108/MIP-06-2022-0266

Fishbein, M., & Ajzen, I. (1975). *Belief, attitude, intention, and behavior: An introduction to theory and research.* Addison-Wesley.

Howell, R. T., Pchelin, P., & Iyer, R. (2012). The preference for experiences over possessions: Measurement and construct validation of the experiential buying tendency Scale. *The Journal of Positive Psychology, 7*(1), 57–71.https://doi.org/10.1080/17439760.2011.626791

Kahneman, D. (2010, February). *The riddle of experience vs. memory.* https://www.ted.com/talks/daniel_kahneman_the_riddle_of_experience_vs_memory

Kotler, P., Pfoertsch, W., & Sponholz, U. (2021). *H2H Marketing.* Springer International Publishing.

Kumar, A., Killingsworth, M. A., & Gilovich, T. (2014). Waiting for merlot: Anticipatory consumption of experiential and material purchases. *Psychological Science, 25*(10), 1924–1931.https://doi.org/10.1177/0956797614546556

Likert, R. (1932). A technique for the measurement of attitudes. *Archives of Psychology, 140*, 1–55.

Maister, D. H., Galford, R., & Green, C. (2021). *The trusted advisor.* Free Press.

Oliver, R. L. (1980). A cognitive model of the antecedents and consequences of satisfaction decisions. *Journal of Marketing Research, 17*(4), 460–469. https://doi.org/10.2307/3150499
Oliver, R. L. (1997). *Satisfaction: A behavioral perspective on the consumer*. McGraw-Hill.
Oliver, R. L. (1999). Whence consumer loyalty? *Journal of Marketing, 63*(Special Issue), 33–44. https://doi.org/10.2307/1252099
Osgood, C. E., Suci, G. J., & Tannenbaum, P. H. (1957). *The measurement of meaning*. University of Illinois Press.
Reichheld, F. F. (2003). The one number you need to grow. *Harvard Business Review, 81*(12), 46–55.
Schmitt, B. H. (1999). *Experiential marketing how to get customers to sense, feel, think*. Free Press, New York.
Stoyanov, D. (2014). Market and marketing peculiarities of vending channels. *IZVESTIA – Journal of University of Economics – Varna, 3*, 74–86.
Stoyanov, D. (2015). Sustainable marketing: A global benchmark perspective on the vending industry. *Review of Integrative Business and Economics Research, 4*(2), 1–19.
Stoyanov, D. (2021a). Marketing of vending channels: A case of French university campuses. *International Journal of Retail & Distribution Management, 49*(5), 577–594. https://doi.org/10.1108/IJRDM-06-2020-0228
Stoyanov, D. (2021b). The role of vending channels in marketing: A systematic review and taxonomy of studies. *The Journal of Consumer Affairs, 55*(2), 654–679. https://doi.org/10.1111/joca.12362
Stoyanov, D. K. (2023). Is the whole greater than the sum of its parts? A vending retail study. *International Journal of Retail & Distribution Management, 51*(3), 327–346.https://doi.org/10.1108/IJRDM-06-2022-0186
Stoyanov, D. K. (2025). *Vending retail marketing: Text and international cases*. Springer Nature, Switzerland
Zajonc, R. B. (1980). Feeling and thinking: Preferences need no inferences. *American Psychologist, 35*(2), 151–175.https://doi.org/10.1037/0003-066X.35.2.151

8 Sustainable Value in Action: Case Studies

8.1 Driving Sustainability in the FMCG Sector: A Case Study[1]

8.1.1 Introduction

In today's dynamic business environment, sustainability has evolved from a niche concept to an indispensable element of new product development and overall business strategy. This shift is driven by a confluence of factors, including stringent government regulations, evolving consumer expectations, the pursuit of operational efficiencies, and profound corporate commitments. As articulated by Jorge Zuniga, Senior Insights Consultant at Euromonitor International, sustainability is no longer merely an "optional extra" but an "essential component" in the creation of new products. It encompasses a dedication to developing products ethically, minimizing environmental impact, and ensuring core benefits such as functionality, flavor, quality, and taste are not compromised. This case study explores the multifaceted landscape of sustainable marketing, drawing insights from recent industry reports and real-world examples.

8.1.2 Key Drivers of Sustainable Innovation

Several powerful forces are propelling companies to integrate sustainability into their core operations and marketing strategies:

[1] This case study is based on recent analyses and reports on sustainability innovations in the FMCG sector of Euromonitor International (Tekutyte, 2024a, 2024b, 2025; Zuniga, 2025; Zuniga & Tekutyte, 2024).

D. K. Stoyanov, *Sustainable Marketing*, Springer Texts in Business and Economics,
https://doi.org/10.1007/978-3-032-16498-8_8

- **Government regulations:** Regulatory bodies worldwide are increasing pressure on businesses regarding climate action. For instance, 71% of global businesses face some form of climate action pressure from regulators. Initiatives like the European Green Deal are actively combating "greenwashing" and advocating for sustainable practices in sourcing, packaging, and waste management. Non-compliance carries significant risks, including financial penalties and reputational damage.
- **Consumer demand:** Consumers are increasingly factoring sustainability into their purchasing decisions. Over 60% of consumers reported trying to have a positive environmental impact in 2024, a trend that has remained consistent since 2020. Furthermore, 65% of global businesses note rising climate action pressure from consumers. Brands that offer credible and accessible sustainable products are more likely to foster consumer loyalty, while those that lag risk losing customers to more committed competitors. Consumer-centric strategies are a high priority, with 60% of companies focusing on customer experience.
- **Financial and operational efficiencies:** Integrating sustainability into innovation can yield substantial operational and financial benefits. Strategies such as optimizing packaging design, improving energy efficiency, and reducing waste can lead to lower costs, reduced carbon footprints, and optimized resource utilization. These practices not only enhance business resilience but also appeal to environmentally conscious consumers and investors seeking ethically responsible and future-ready businesses.
- **Corporate commitments:** Sustainability is becoming a central tenet of corporate strategies, with 61% of global companies adopting sustainable practices. Consumer goods brands are setting ambitious targets to reduce carbon emissions, decrease reliance on virgin materials, and minimize waste, all while promoting ethical operations. Achieving these objectives necessitates innovative solutions and transparent progress tracking, making sustainability a significant competitive advantage that attracts investors, top talent, and consumer loyalty.

8.1.3 The Evolving Consumer Landscape and Purchase Barriers

While consumer awareness and concern for environmental issues are high, certain barriers hinder widespread sustainable purchasing. The primary obstacle, cited by 40% of global respondents in a January–February 2024 survey, is the high price of sustainable alternatives compared to conventional products. Other significant barriers include unclear labeling, uncertainty about identifying sustainable features, limited availability of sustainable products in typical shopping locations, and a lack of trust in sustainable product claims.

The "Consumers' Awareness and Action 2020–2024" data indicates that while over 60% of global respondents consistently worried about climate change and tried to have a positive impact, the percentage who felt they could make a difference remained lower and relatively stable during the same period. This highlights a potential gap between concern and perceived efficacy of individual action. To overcome these barriers, businesses must prioritize clear, transparent communication that links green attributes directly to core product benefits such as efficacy, quality, or safety. Consumers are adopting a more practical approach, favoring sustainable options that offer comparable benefits at an affordable price.

8.1.4 Innovation in Sustainability Across the FMCG Supply Chain

Sustainability is being embedded across all stages of the fast-moving consumer goods (FMCG) supply chain, from sourcing to retail. As of 2023, 47% of global companies incorporated sustainability features into new product launches, and 58% plan to develop sustainable products within the next five years. Since 2023, 27,234 new FMCG products have been launched with at least one sustainability claim. Sustainable products are also outperforming non-sustainable ones in terms of retail sales compound annual growth rate (CAGR) in nine out of eleven industries between 2020 and 2023.

Sourcing

Companies are increasingly pressured to demonstrate transparent carbon reduction strategies, leading to innovations in ethical and environmentally responsible sourcing. This includes re-evaluating product formulations to favor cleaner, more natural ingredients with minimal environmental and community impact. Key areas of innovation include:

- **Regenerative agriculture:** Companies are supporting farmers in adopting practices that improve soil health, sequester carbon, and enhance crop resilience. For example, Cascadian Farm's limited edition Climate Smart Kernza Grains cereal uses a perennial grain that helps improve soil health and reduce water pollution.
- **Novel technologies:** Investments are being made in technologies like cultured meat and precision fermentation to create more sustainable supply chains. BettaF!sh, for instance, offers a vegan salmon made from organic European seaweed and plant proteins, leveraging zero-input seaweed cultivation that absorbs CO2 and boosts underwater biodiversity.
- **Waste reduction and alternative ingredients:** Brands like Almond Cow provide products and organic nuts and fruits that enable consumers to make plant-based milk at home, reducing waste from pre-packaged milk. Valio's Oddlygood plant-based product line emphasizes sustainable practices and traceable sourcing.
- **Ethical supply chains:** The snack industry, particularly with ingredients like chocolate, has faced challenges related to deforestation and unfair labor practices. Companies like Tony's Chocolonely are gaining traction with their 5-step sustainable sourcing practices, which include paying higher wages and ensuring

100% traceable beans. Precision fermentation is also being explored to replace cocoa with cereals and legumes, resulting in lower emissions.

Overall, 73% of global companies are either investing in sustainable sourcing or plan to do so in the near future. The global retail value sales of products carrying at least one sustainable sourcing claim reached USD 196 billion in 2023. "Organic" is the most adopted claim, present in seven out of ten new dairy products and alternatives launched with sustainable sourcing features in 2023.

Manufacturing

Innovation in sustainable manufacturing is closely tied to corporate net-zero commitments, moving beyond mere carbon neutrality to genuine reductions in operational emissions. This involves:

- **Zero waste practices:** Companies are adopting strategies to minimize waste directed to landfills and optimize resource utilization.
- **Responsible water management and energy efficiency:** These are common priorities in sustainable manufacturing.
- **Product reformulation:** Firms are reformulating products to exclude harmful chemicals and substituting animal-based ingredients with greener alternatives, such as upcycled components. Such changes can offer carbon-reducing advantages and cost efficiencies. Upcycled products are gaining traction in premium fragrances and facial care.
- **Transparency and certification:** Due to skepticism surrounding carbon offset projects, brands are increasingly relying on third-party certifications to demonstrate accountability for their carbon reduction claims.

Packaging

Sustainable packaging design is crucial for reducing environmental impact, focusing on minimizing waste and optimizing resource use. Innovations in this area aim to decrease the carbon footprint and overall costs by reducing reliance on virgin materials. Companies are exploring various approaches to make packaging more eco-friendly, including:

- **Minimizing material use:** Designing packaging to use less material overall, reducing both production impacts and waste.
- **Reusable and refillable solutions:** Implementing systems where packaging can be reused multiple times or refilled, significantly cutting down on single-use waste.
- **Recyclable and compostable materials:** Shifting toward materials that can be easily recycled or composted at the end of their life cycle, diverting waste from landfills.
- **Alternative packaging materials:** Developing and utilizing innovative materials like agricultural waste, mycelium, or seaweed-based polymers that offer

sustainable alternatives to traditional plastics. For example, brands are exploring edible packaging or packaging made from renewable resources.
- **Smart packaging:** Integrating technologies that provide consumers with information on how to properly dispose of or recycle packaging, improving consumer engagement in the circular economy.

Logistics/Distribution

Optimizing logistics and distribution networks is critical for reducing the environmental footprint of product delivery. Innovations in this area focus on efficiency, route optimization, and the adoption of greener transportation methods:

- **Route optimization:** Utilizing advanced software and data analytics to plan the most efficient delivery routes, minimizing fuel consumption and emissions. This includes consolidating shipments and optimizing vehicle load capacities.
- **Alternative fuels and electric vehicles (EVs):** Investing in fleets that run on alternative fuels (e.g., biofuels, hydrogen) or transitioning to electric vehicles for last-mile delivery, significantly reducing greenhouse gas emissions and air pollution in urban areas. For example, some companies are deploying electric vans or bikes for urban deliveries.
- **Warehouse efficiency:** Implementing sustainable practices within warehouses, such as using renewable energy sources, optimizing lighting and temperature control, and improving waste management, further reducing the overall environmental impact of logistics operations.
- **Circular logistics:** Designing reverse logistics systems for collecting used products or packaging for reuse, recycling, or proper disposal, supporting a more circular economy.
- **Collaborative logistics:** Partnering with other companies to share transportation resources and routes, reducing the number of half-empty trucks on the road and improving overall efficiency.

Retail

Retailers play a pivotal role in promoting sustainability across the supply chain by directly influencing consumer choices and implementing sustainable practices within their operations. Their involvement includes:

- **Partnerships with sustainable suppliers:** Collaborating with brands and manufacturers that prioritize sustainable sourcing, ethical labor practices, and environmentally friendly production methods, ensuring that the products on their shelves align with sustainability goals.
- **Making eco-friendly shopping accessible:** Offering user-friendly tools and clear labeling to help consumers easily identify and choose green products. This can include dedicated eco-sections in stores, online filters for sustainable attributes, and educational campaigns.

- **In-store sustainability initiatives:** Investing in renewable energy sources for stores and distribution centers, implementing efficient lighting and HVAC systems, and adopting comprehensive waste reduction and recycling programs within their retail spaces.
- **Sustainable private label ranges:** Developing their own lines of sustainable products, which often allows for greater control over the supply chain and provides more affordable eco-friendly options for consumers.
- **Consumer engagement and education:** Actively engaging with consumers through marketing campaigns that highlight the benefits of sustainable products and provide information on how consumers can contribute to environmental efforts. This also involves encouraging practices like recycling and bulk purchasing.
- **Promoting circularity:** Facilitating product take-back programs, repair services, and refill stations to extend product life cycles and reduce waste.

8.1.5 Discussion Questions

1. Given the identified barriers to sustainable purchases, particularly high prices and unclear labeling, what innovative marketing strategies can companies employ to make sustainable products more appealing and accessible to the average consumer without compromising profitability?
2. The reports highlight the importance of "authentic operational emissions reduction" over "carbon neutrality" in manufacturing. How can a company effectively communicate its genuine efforts in reducing emissions to skeptical consumers and avoid accusations of greenwashing?
3. Considering the diverse innovations across the supply chain (sourcing, manufacturing, packaging, logistics, retail), which stage presents the most significant opportunity for sustainable marketing to influence consumer perception and drive purchase decisions, and why?
4. The case study notes that while many consumers are worried about climate change, fewer feel they can make a difference. How can sustainable marketing campaigns empower consumers and bridge this gap between concern and action?
5. With the rise of novel technologies like cultured meat and precision fermentation in sourcing, what ethical and marketing challenges might arise for companies adopting these innovations, and how can they be addressed proactively?

8.2 Sustainable Communications in the Fragrance Industry: A Case Study[2]

8.2.1 Overview

Sustainability is reshaping the global fragrance industry. Once a peripheral concern, it is now embedded at the core of product development, branding, and consumer communication strategies. This case study invites students to explore how sustainability is marketed in one of the most sensorial and emotionally charged product categories: fragrances. Between 2021 and 2023, the market for sustainable fragrances expanded rapidly, reaching over USD 10 billion in value. Western Europe, in particular, has emerged as a global leader in this space, home to five of the world's ten largest sustainable fragrance markets. Yet, this transformation brings challenges—ranging from consumer skepticism and high product costs to regulatory scrutiny and marketing complexity. How can fragrance brands communicate sustainability authentically and effectively? What messaging tools and platforms resonate best with eco-conscious consumers? This case explores the answers to these questions through recent trends and examples from industry leaders.

8.2.2 The Rise of Sustainability Claims

The dominant trend in fragrance marketing is the shift from conventional sensory appeals toward ethically framed messaging. Leading claims include:

- "Natural," which, despite its popularity, remains poorly defined and potentially misleading without proper substantiation.
- "Vegan," "Cruelty-free," and "Environmentally friendly," which have grown sharply in consumer visibility and SKU count globally.
- "From recycled materials," a packaging-focused claim that has gained traction with brands like Givenchy and LVMH, especially in high-end segments.

The trend toward upcycled ingredients—such as cedarwood essence derived from sawdust—is reshaping consumer expectations. Brands like Carolina Herrera and La Bouche Rouge have capitalized on this innovation, transforming waste into luxury through compelling storytelling.

[2] This case study is based on a recent report on sustainability communications in the fragrance industry from Euromonitor International (Rajpara, 2024).

8.2.3 Packaging as Communication

Sustainable packaging is not just a logistics concern-it has become a core narrative element. In 2023, fragrance SKUs featuring packaging claims such as "recyclable," "plastic-free," and "refillable" saw a major increase. Over 75% of companies in the beauty and personal care industry now cite packaging as a central pillar of their sustainability strategy. The use of materials like aluminum (e.g., by INCA Packaging) offers dual benefits: environmental compatibility and a premium appearance. This aligns well with consumer preferences in Western Europe, where refillable bottles and recyclable containers are seen as key differentiators.

8.2.4 The Trust Gap and Regulatory Shift

Despite the growth of sustainability claims, consumer trust remains a barrier. Only 44% of Europeans said they trust labels like "No Artificial Ingredients." Trust issues are particularly prevalent among older consumers, who report difficulty interpreting vague or non-standardized claims. To address this, the European Union introduced the Green Claims Directive in 2023, requiring companies to provide robust evidence for any environmental claim made on packaging or in advertising. In parallel, the UK launched anti-greenwashing regulations under its Financial Conduct Authority. These measures are pushing brands to adopt third-party certifications, such as Cradle to Cradle, and to substantiate claims using life cycle assessments.

8.2.5 Innovation in Communication Channels

Traditional sustainability communication—via websites and financial reports—has limited impact on purchase decisions. Only 27% of companies believe their sustainability messaging is effective. However, there is growing momentum for packaging-based communication tools like QR codes. A notable example is Bastille Parfums, a French niche brand that embedded QR codes on product bottles to share information about sourcing, formulation, and environmental impact. This innovation appeals especially to younger consumers who demand transparency and interactivity.

8.2.6 Regional Differences and Market Strategy

Consumer preferences are **not uniform across markets**:

- In France, demand has shifted toward eco-friendly packaging, with refillable bottles gaining popularity.

- In Spain and Italy, emerging consumer awareness has sparked interest in ingredient transparency and ethical sourcing.
- In Germany and the UK, sustainability is viewed as standard, not premium—consumers expect brands to provide it without passing on high costs.

Meanwhile, markets like Asia Pacific and the Middle East remain underserved due to lower consumer awareness and distribution challenges. However, these regions represent high-growth opportunities for brands willing to invest in education and infrastructure.

8.2.7 Discussion Questions

1. How can fragrance brands design sustainability claims that are both emotionally engaging and verifiable under tightening regulations? Propose a communication strategy for a mid-sized European brand.
2. Evaluate the risks and benefits of positioning sustainability as a luxury feature in the fragrance market. How does price perception influence the effectiveness of such messaging?
3. What role can product packaging play in enhancing transparency and consumer trust in sustainability claims? How could smart packaging tools like QR codes be better utilized in mass markets?
4. Given the differences across European countries, how should brands tailor their sustainable communication strategies by region? Suggest a messaging adaptation for Spain versus France.
5. Imagine you are launching a fragrance made entirely from upcycled ingredients. Outline a cross-channel marketing plan that balances emotional storytelling with credible, data-backed claims.

8.3 Trends in Sustainable Retailing: A Case Study[3]

8.3.1 Introduction

In the contemporary retail landscape, sustainability has transcended its former status as a niche concern to become a central imperative, driven by evolving consumer expectations and increasingly stringent governmental regulations. Retailers globally are recognizing the critical need for transformative action, particularly in the realms of climate-positive initiatives and the transition toward a circular economy. This case study examines the diverse approaches adopted by six prominent retailers—Amazon.com, Inc., Decathlon SA, Fast Retailing Ltd (FRC), Natura &

[3] This case study is based on a recent report of sustainability trends in the retailing industry from Euromonitor International (Milasevic, 2024).

Co, Recreational Equipment Inc (REI), and Vinted UAB—as they navigate the complexities of integrating sustainability into their core business strategies. The urgency for change is palpable. Nearly half of global consumers actively engage in efforts to combat climate change, fostering an expectation that retailers will mirror their commitment. While consumers advocate for environmental responsibility, their willingness to pay a premium for sustainable products remains limited, especially amidst economic pressures like the cost-of-living crisis. This creates a challenging dynamic where retailers are expected to embed sustainability as a standard offering without significantly increasing costs to the consumer. For instance, in 2023, only 9% of global consumers were prepared to pay more for eco-friendly apparel or footwear, highlighting a significant affordability barrier. Technology plays a crucial role in advancing retail sustainability, enabling greater efficiency, fostering innovation, and promoting environmentally sound practices. From optimizing supply chains to integrating renewable energy sources, technological solutions are vital for achieving a sustainable future. Furthermore, the burgeoning second-hand market is a key driver of the circular economy, with retailers actively participating in resale initiatives to meet sustainability targets and capitalize on this growing segment. In 2023, approximately 24% of digital consumers globally planned to increase their spending on pre-owned items in the coming year, reflecting a growing trend.

Despite these advancements, retailers face considerable hurdles, including fragmented efforts and a lack of unified vision or clear measurement frameworks for sustainability initiatives. Regulatory influence, particularly from the European Union's leading environmental policies, is anticipated to be a significant catalyst for progress, compelling greater accountability and more structured reporting. While retailers are intensifying efforts to minimize their environmental impact through recycling and energy efficiency, managing Scope 3 emissions—those indirectly generated across their value chains—remains a persistent challenge due to reliance on external partners. Effectively addressing these emissions necessitates a holistic approach that extends beyond internal operations to encompass partner practices. This case study delves into the specific strategies, successes, and ongoing challenges faced by these six retailers, providing a rich context for understanding the practical application of sustainable marketing principles in a dynamic global market.

8.3.2 Retailer Profiles and Sustainability Strategies

Amazon.com Inc: Amazon, a global e-commerce and technology giant with global retail sales of USD 528.8 billion in 2022, has actively pursued sustainability despite its business model often being associated with overconsumption. As a co-founder of the Climate Pledge in 2019, Amazon has influenced approximately 400 companies to commit to achieving carbon neutrality by 2040. In 2022, 90% of Amazon's electricity consumption was sourced from renewable energy, and its carbon intensity decreased by 7% from 2021 to 2022. The company's circular economy efforts include the Amazon Second Chance program, launched in 2018, which facilitates recycling of packaging, product repairs (over 7.6 million products repaired in 2022), and trade-ins. Amazon also partners with organizations like the Ellen MacArthur Foundation to promote circularity, including developing reusable packaging solutions. However, Amazon's total emissions increased by 40% since 2019, despite a slight 0.4% decrease between 2021 and 2022, indicating ongoing challenges in decarbonizing its extensive operations, particularly direct emissions which rose by 11% in that period.

Decathlon SA: Decathlon, the world's leading sports goods retailer with global retail sales of USD 15.9 billion in 2022, operates over 1,700 stores across more than 70 countries. Its "2020–2026 Transition Plan" emphasizes CO2 emission reduction, eco-designed products (23% of products in 2022), and scaling circular business models. In 2022, 11% of its product catalog was qualified as repairable. Decathlon promotes circularity through initiatives like its "Nolhtaced" logo reversal in Belgium stores to encourage buy-back services, collecting nearly 160,000 items in 2022. The company has achieved a reduction in absolute carbon emissions for the first time in 2022 while maintaining sales growth. Despite 95% of packaging materials being from sustainable sources, packaging and product transportation remain areas for improvement in greenhouse gas emissions.

Fast Retailing Ltd (FRC): Fast Retailing, the parent company of UNIQLO and other apparel brands, reported global retail sales of USD 15.0 billion in 2022. Its "LifeWear" concept aims for sustainability through durable, high-quality clothing, sustainable production, and recycling. FRC's Fiscal 2030 Sustainability Targets include a 90% reduction in greenhouse gas emissions from its own operations and 20% from its supply chain, along with increasing recycled materials in products to 50% (currently 5% in 2022). Programs like RE.UNIQLO and RE.GU collected 16.7 million second-hand clothing items in fiscal 2022 for donation or recycling. While FRC reduced absolute emissions in its own operations by 46% from fiscal 2019 to 2022, supply chain emissions remain substantial.

Natura & Co: Natura & Co, a Brazilian beauty and cosmetics company with global retail sales of USD 10.7 billion in 2022, has been carbon-neutral since 2007. Its "Commitment to Life" strategy focuses on addressing the climate crisis, protecting the Amazon (preserving over 2 million hectares), defending human rights, and embracing circularity. The company aims for complete packaging circularity, with 82% of materials designed to be reusable, recyclable, or compostable. Programs like "Recycle with Natura" in Brazil encourage packaging returns for discounts, and in India, 617 tons of plastic waste were acquired for recycled packaging in 2022. The Body Shop, a Natura & Co brand, has over 800 refill stations worldwide, preventing the disposal of 605,000 plastic bottles. Natura & Co's 2021 Impact & Profit & Loss (IP&L) program indicated a USD 1.5 positive societal impact for every USD 1 in revenue.

Recreational Equipment Inc (REI): REI, a US-based co-op specializing in outdoor gear and apparel with global retail sales of USD 3.9 billion in 2022, strongly emphasizes sustainability. Since 2020, production and delivery of REI-branded products have been carbon neutral, and its stores, distribution centers, and offices use 100% renewable energy. REI's circular economy initiatives are robust, including rentals, in-store repairs (over 300,000 items repaired in 2022), and the Re/Supply trade-in program, which sells over one million used items annually and bought back over 100,000 used products in 2022. Despite these efforts, REI's total carbon footprint, including emissions from partner brands, increased by 7% from 2019 to 2022, with over 52% of its 2022 emissions linked to third-party brands. REI aims to reduce its total carbon footprint by 55% by 2030 and plans for 100% of products sold to have preferred sustainability attributes by the same year.

Vinted UAB: Vinted, a Lithuanian-based online C2C marketplace for second-hand fashion, reported global retail sales of USD 405 million in 2022, with a 51% year-on-year increase. Its business model inherently promotes circularity, aiming to make second-hand shopping a first choice globally. Vinted's Climate Impact Report revealed a net reduction of approximately 453 kilotons in carbon emissions in 2022 due to its marketplace activities. The company focuses on decarbonizing logistics, with 73% of transactions utilizing Pick-Up and Drop-Off (PUDO) points, reducing emissions by 62% compared to home delivery. Additionally, 62% of sellers reused single-use packaging, leading to 70% lower emissions than new packaging. While Vinted's growth and focus on affordability and convenience drive sustainable consumer behavior, its recent expansion to include small businesses selling new products could potentially dilute its eco-conscious image.

8.3.3 Key Takeaways for Sustainable Marketing

The strategies of these leading retailers underscore several critical lessons for sustainable marketing. Firstly, environmental regulations, particularly from regions like the EU, are increasingly shaping corporate sustainability agendas, mandating greater transparency and reporting. Secondly, the circular economy, extending beyond just the second-hand market to include renting, sharing, repairing, and reusing, is a powerful engine for innovation and long-term value creation in retail. Thirdly, while climate concerns resonate with both businesses and consumers, a significant gap exists between consumer desire for sustainable options and their willingness to pay more for them, especially in challenging economic climates. This necessitates retailers to integrate sustainability as a standard, cost-neutral offering.

Furthermore, effective sustainable marketing requires addressing Scope 3 emissions by collaborating with supply chain partners, as not all environmental impacts are within a retailer's direct control. Retailers must also prioritize educating and engaging consumers about their sustainability initiatives and how consumers can actively contribute. Finally, developing comprehensive sustainability strategies with clear metrics and accountability measures is paramount to ensure genuine progress and combat greenwashing. Despite a recognized opportunity for improvement in communicating sustainability efforts, new regulations and reporting requirements are expected to foster greater trust in product sustainability claims, emphasizing the importance of transparent and credible communication.

8.3.4 Discussion Questions

1. Given that nearly half of global consumers actively work to address climate change, yet only a small percentage (e.g., 9% for apparel/footwear in 2023) are willing to pay more for sustainable products, how can retailers effectively bridge this "intention-action gap" in their sustainable marketing strategies? What value propositions, beyond just environmental benefits, can compel consumers to choose sustainable options without significantly increasing price sensitivity?
2. Decathlon and REI have demonstrated successful circular economy initiatives through buy-back programs, repairs, and rentals. However, challenges remain, such as the perceived lack of sustainability in apparel and footwear production (Decathlon) or the impact of third-party brands on overall carbon footprint (REI). Discuss the primary challenges retailers face in fully transitioning to a circular economy model, particularly concerning product design, supply chain integration, and consumer adoption, and propose innovative marketing strategies to overcome these hurdles.
3. Amazon and Fast Retailing highlight the significant challenge of Scope 3 emissions, which largely originate from activities outside their direct control but within their value chains. How can retailers leverage their market influence and sustainable marketing principles to effectively engage and incentivize their supply chain partners to reduce Scope 3 emissions? What role can transparency and accountability play in driving this collaborative effort?
4. The case study mentions the anticipation of increased trust in product sustainability claims due to new regulations and reporting requirements. However, greenwashing remains a concern. How can retailers ensure the credibility and authenticity of their sustainable marketing messages, especially when their business models (like Amazon's e-commerce or Vinted's expansion to new items) might inherently pose sustainability challenges? What specific communication strategies and third-party certifications can enhance consumer trust?
5. Technology is identified as pivotal in driving sustainability solutions, from smart supply chains to resale platforms. Analyze how retailers can best utilize data and technological advancements not only for internal efficiency and emission reduction but also to inform and enhance their sustainable marketing efforts. Consider how data can be used to communicate impact, personalize sustainable offerings, and foster deeper consumer engagement.

8.4 The Tobacco Industry's Sustainable Shift—A Marketing Challenge: A Case Study[4]

8.4.1 Introduction

The global tobacco industry, valued at approximately USD 931 billion, faces an unprecedented challenge: reconciling its historical product offerings with the growing global imperative for sustainability. While cigarettes still dominate the market, accounting for 82% of the industry's value, newer categories like e-vapor products, heated tobacco, smokeless tobacco, and oral nicotine are gaining traction. This shift is not merely about product innovation; it's a complex response to escalating pressure from consumers, investors, and regulators who demand greater accountability for environmental, social, and governance (ESG) impacts. This case study explores the multifaceted sustainability journey of the tobacco industry, highlighting the inherent tensions and strategic dilemmas in its pursuit of a more sustainable future.

8.4.2 The Shifting Landscape: Pressures and Paradoxes

The drive for sustainability in the tobacco sector is a "double-edged sword," presenting both significant threats and novel opportunities. The industry is grappling with declining cigarette smoking prevalence in many markets, partly influenced by changing mobility patterns and public health concerns exacerbated by events like the COVID-19 pandemic. This has spurred a broadening of nicotine portfolios toward reduced-risk products (RRPs) to retain consumer bases. Simultaneously, regulatory innovations, such as New Zealand's "birth year ban" aimed at achieving smoke-free goals, are intertwining with the adoption of sustainability measures. This challenging environment is pushing tobacco companies not only toward next-generation products but also to explore categories beyond nicotine, such as cannabis, healthcare, and biotechnology.

8.4.3 Consumer Dynamics: Price vs. Planet

Consumers are increasingly environmentally aware, with popular actions including reducing plastic use and recycling. For instance, in 2023, these actions were prioritized over directly climate-linked behaviors like reducing energy consumption. This trend presents an opportunity for tobacco companies to innovate, for example, by replacing plastic and acetate tow filters in traditional cigarettes with more eco-friendly alternatives. Some major players, like BAT, report that 93% of their

[4] This case study is based on a recent report of sustainability trends in the tobacco industry from Euromonitor International (Becerril & MacGuill, 2023).

cigarette plastic packaging is recyclable, and Japan Tobacco aims for fully recyclable packaging by 2030. However, a significant paradox emerges: while many tobacco users express a commitment to green activities, they are often reluctant to accept sustainability initiatives that might lead to higher prices or alter their current product experience. Affordability and convenience remain critical drivers of consumer behavior. The rise of disposable e-vapor products, often cheaper than combustible cigarettes, illustrates this point. Their success, driven by ease of access and affordability rather than sustainability, raises questions about consumers' actual environmental commitment. This price sensitivity is a major barrier, with a 2023 survey indicating that a lack of consumer willingness to pay more for sustainable products is the most significant obstacle to corporate sustainability initiatives. In countries with high excise taxes, the ability of tobacco companies to pass on compliance costs is limited, risking reduced demand or an increase in illicit trade. Furthermore, communicating sustainability efforts is challenging due to strict advertising and promotion restrictions. While product claims and packaging are growing as communication tools, graphic warnings and plain packaging laws in many countries (e.g., Australia, France, the UK) severely limit space for such messages. This forces companies to prioritize markets without plain packaging, such as Germany, for launching new products with sustainable features like paper filters.

8.4.4 Investor Influence: Engagement vs. Divestment

ESG considerations are becoming a cornerstone of investment decisions. A PwC report projected that ESG-linked assets under management would reach USD 34 trillion by 2026, up from USD 18 trillion in 2021. Investors increasingly view strong ESG performance as an indicator of a company's overall resilience and long-term viability. While historical investor activism often involved divestment from tobacco stocks, there's a growing sentiment among some investors that engagement, coupled with ESG action and reporting, can accelerate the industry's transformation. However, the question "Can tobacco companies ever be sustainable?" remains contentious. Critics argue that sustainability efforts are merely "greenwashing" to distract from the inherent harm of their products. Others contend that sustainability must be viewed through the lens of transitioning away from traditional, harmful nicotine consumption toward safer alternatives. This transformation itself presents new sustainability challenges, particularly regarding the impact on tobacco farmers due to reduced leaf demand and the environmental footprint of electronic RRPs (e.g., plastic, copper, and lithium waste from disposable vapes). The recent surge in disposable vape use has starkly highlighted this electronic waste issue. The misalignment in value perception between ESG-focused investors and those prioritizing short-term returns creates a dilemma. This could lead to tobacco companies spinning off their less controversial nicotine businesses to attract ESG-conscious investors, leaving legacy combustible-only operations to a narrower, potentially private, investor base.

8.4.5 Regulatory Imperatives: From Product Waste to Supply Chains

Regulators are playing an increasingly assertive role in pushing for sustainability. Key areas of focus include reducing product waste, particularly cigarette butts and electronic waste from RRPs, and addressing environmental impacts across the supply chain, from leaf cultivation to manufacturing emissions. The World Health Organization (WHO) estimates that 4.5 trillion cigarette butts are improperly discarded annually, making them the largest type of litter globally. In the UK alone, an estimated two disposable vapes are discarded every second. Legislation like the EU Single-Use Plastics Directive (2021) places extended producer responsibility on cigarette manufacturers for discarded butts, a model likely to be replicated elsewhere. This directive may also impact e-vapor products, pushing for more environmentally friendly materials. Some tobacco companies, including BAT, Imperial Brands, and Landewyck, are testing paper filters as alternatives to plastic and acetate tow in anticipation of potential bans. Beyond product waste, regulators are scrutinizing supply chain emissions and leaf cultivation practices, which contribute to deforestation, soil degradation, and water usage. The WHO's 2022 report, "Tobacco: Poisoning our Planet," highlights that tobacco farming accounts for about 5% of global deforestation and that 22 billion tons of water are used annually in global tobacco production. While major tobacco companies are self-regulating and setting ambitious targets (e.g., net-zero carbon emissions by 2050 for their entire value chain), due diligence laws, such as those proposed by the European Commission, are expected to increase scrutiny, making companies liable for violations within their immediate supply chains.

8.4.6 Industry Response: Innovation and Communication Barriers

The tobacco industry is responding with significant investments in RRPs and sustainability initiatives. Companies like Philip Morris International (PMI) aim for smoke-free products to account for over 50% of their revenues by 2025, while BAT targets GBP 5 billion in revenue from "New Categories" by the same year. This shift involves developing complex recycling schemes for RRPs, with some companies implementing "reward" schemes to boost recycling rates for disposable products. PMI has even opened an e-vapor recycling center in Budapest, capable of processing 150,000 devices monthly. However, the industry faces the challenge of consumer acceptance for new, more sustainable product features (e.g., paper filters) that might alter the smoking experience. Past attempts at non-plastic filters had limited consumer uptake. Furthermore, despite ambitious environmental goals and significant investments, strict regulations on promotion and communication, particularly plain packaging laws, hinder the industry's ability to effectively communicate its sustainability efforts to stakeholders. This creates a marketing conundrum: how can an industry with a controversial core product effectively convey its commitment to sustainability when its communication channels are severely restricted?

8.4.7 Discussion Questions

1. Given the inherent health risks associated with tobacco products, can the tobacco industry genuinely achieve "sustainability" in a meaningful sense, or are its efforts primarily perceived as "greenwashing" by consumers, investors, and regulators? What are the key ethical considerations for a sustainable marketing professional working in this industry?
2. Consumers express a desire for sustainable products but are often unwilling to pay more or sacrifice convenience. How can tobacco companies effectively market and incentivize the adoption of more sustainable, yet potentially more expensive or less convenient, reduced-risk products and features (e.g., paper filters, recycling schemes) without driving consumers toward illicit markets or cheaper, less sustainable alternatives?
3. The tobacco industry faces severe restrictions on advertising and promotion. How can companies creatively and ethically communicate their sustainability initiatives and product innovations to stakeholders within this highly regulated environment? What alternative communication channels or strategies could be most effective?
4. The case highlights a potential misalignment between ESG-focused investors and those prioritizing traditional returns. If you were a sustainability marketing lead for a major tobacco company, how would you craft a communication strategy to attract and retain both types of investors, articulating the long-term value proposition of sustainability initiatives while acknowledging the financial realities?
5. The shift to RRPs impacts tobacco farmers and creates new waste streams. Beyond product-focused sustainability, what broader social responsibility initiatives should tobacco companies prioritize to ensure a just transition for their supply chain partners and to manage the environmental impact of electronic waste? How can these efforts be integrated into a cohesive sustainable marketing narrative?

8.5 Understanding the Modern Sustainable Consumer: A Case Study[5]

8.5.1 Introduction

In an era of increasing environmental awareness and social responsibility, businesses are navigating a complex landscape where consumer values profoundly influence purchasing decisions. The "Voice of the Consumer: Sustainability Survey 2024 Key Insights" report by Euromonitor International provides crucial insights into how consumers worldwide perceive and act on sustainability. This case study examines key findings from the report, highlighting the diverse motivations and behaviors of various "Sustainable Consumer" segments. It aims to equip students with a practical understanding of sustainable marketing principles, encouraging critical thinking on how businesses can effectively engage with these evolving consumer types. The survey, conducted annually in January and February, gathers responses from 1,000 online consumers aged 15 to 74 across 39 countries, focusing on attitudes, expectations, and adoption related to sustainability. It identifies distinct consumer types based on their specific values and priorities concerning environmental issues. While climate change concern is a shared trait among these segments, their engagement with sustainability varies significantly. Businesses must recognize that sustainable attributes often serve as valuable complements rather than primary purchase motivators, emphasizing the need to integrate these attributes with a product's core value proposition.

8.5.2 Key Sustainable Consumer Segments and Their Behaviors

The report identifies several prominent sustainable consumer types, each with unique characteristics and preferences that demand tailored marketing approaches.

Meat Avoiders: This segment, representing over 36% of global consumers, actively reduces or eliminates meat consumption due to environmental concerns or a desire for a more sustainable lifestyle. Millennials constitute the largest portion of Meat Avoiders at 40%, and 42% of them hold graduate or postgraduate qualifications. Geographically, India, Vietnam, and Brazil show the highest percentages of Meat Avoiders, with 52%, 47%, and 47% respectively. Companies can appeal to this group by highlighting "vegan," "vegetarian," and "plant-based" claims, particularly in food and beauty sectors, and by providing transparent information on ingredient origins and environmental impact. Notably, 86% of Meat Avoiders trust "vegan" claims, 76% trust "recyclable" claims, and 72% trust "locally sourced"

[5] This case study is based on a recent report of sustainability trends in consumer behavior from Euromonitor International (Zuniga, 2024).

claims. A significant barrier to sustainable purchases for this group is high price, cited by 46% of consumers, and unclear labeling, cited by 33%.

Zero Wasters: Approximately one-third of consumers globally are identified as Zero Wasters, focusing on recycling, repairing items rather than replacing them, and purchasing second-hand goods. Baby Boomers account for the highest global percentage of Zero Wasters at 39%. This group is also notably present in the Philippines, with 50% of consumers identifying as Zero Wasters. Economic factors heavily influence their behavior, with recycling being a cost-free practice adopted by half of global consumers. The appeal of second-hand products, which offer good quality at lower prices, resonates particularly with younger, price-sensitive consumers. Similar to Meat Avoiders, high price (50%) and unclear labeling (36%) are significant barriers to sustainable purchases for Zero Wasters.

Climate Activists: Around one-third of global consumers are Climate Activists, characterized by their efforts to reduce energy consumption, use energy-efficient products, and decrease carbon emissions through transportation choices or carbon offsetting. Baby Boomers represent the largest share of Climate Activists at 36%. India leads with 44% of its consumers being Climate Activists. These consumers balance sustainability with economic considerations and require transparent evidence of product efficiencies, including cost and energy savings. Skepticism around carbon offsetting projects means companies must validate environmental commitments with credible data. For instance, only 1.2% of products on digital shelves in Sweden carry carbon-neutral or reduced claims due to such skepticism. Price (50%) and unclear labeling (36%) remain key purchasing barriers for this group.

Green Spenders: Over a quarter of global consumers are Green Spenders, defined by their willingness to pay more for sustainable attributes. Millennials form the largest subgroup within Green Spenders, at 30%. Globally, 33% of consumers with annual household incomes over USD150,001 are Green Spenders. India has the highest percentage of Green Spenders at 51%, especially within food and beverages. These consumers view sustainable features as added value and scrutinize the entire product life cycle. They prioritize clear sustainability features like recyclability or sustainably sourced ingredients. High price (47%) and unclear labeling (38%) are notable barriers to their sustainable purchases.

Water Savers: This segment constitutes one-fifth of global consumers and focuses on reducing water usage to positively impact the environment. Baby Boomers are the largest group of Water Savers at 26%, demonstrating more conscious water use than younger generations. Colombia has the highest percentage of Water Savers, at 37%. Water Savers are willing to pay more for eco-friendly and water-efficient products. They are particularly prevalent in Latin America, which saw the most significant increase in per capita expenditure on water and miscellaneous domestic services between 2020 and 2024, and Europe, known for its high per capita consumer spending. Companies can engage them by showcasing water-saving technologies in manufacturing and a commitment to reducing carbon footprints, especially in water-intensive industries and stressed regions. High price (53%) and unclear labeling (37%) are significant barriers for this group.

8.5.3 Discussion Questions

1. Discuss how a global food manufacturer could effectively segment and target "Meat Avoiders" across different regions, considering their varying cultural and economic contexts (e.g., India vs. Europe). What specific product claims (e.g., "vegan," "plant-based") and communication strategies would be most impactful, given that 86% of Meat Avoiders trust "vegan" claims?
2. "Zero Wasters" prioritize recycling and repairing items, with 39% of Baby Boomers globally falling into this category. How can a major apparel brand redesign its product life cycle and marketing messages to appeal to both older generations (Baby Boomers) and younger generations (Gen Z, who prioritize second-hand shopping for affordability and style) within the "Zero Wasters" segment? Consider the "Right to Repair" movement gaining traction in Europe and the US, and how businesses can implement initiatives like resale-as-a-service models.
3. "Climate Activists" seek transparent proof of corporate climate action, and 50% cite high prices as a barrier to sustainable purchases. Discuss the challenges and opportunities for an electronics company in communicating its carbon reduction strategies. How can it overcome consumer skepticism toward "greenwashing" and build trust, especially given that only 1.2% of products on digital shelves in Sweden carry carbon-neutral or reduced claims? What role do certifications and verifiable data play in convincing these consumers?
4. "Green Spenders" are willing to pay more for sustainable attributes, but not at the expense of quality or performance, with 47% noting high price as a barrier. How can a premium beauty brand justify a higher price point for a sustainably sourced product to "Green Spenders"? Explore strategies for incorporating multiple sustainability features and effectively communicating these benefits to demonstrate added value and foster loyalty among these highly educated and environmentally aware consumers.
5. "Water Savers" are a significant consumer group, with 26% of Baby Boomers identifying as such, and 53% of them cite high price as a barrier to sustainable purchases. Analyze how companies in water-intensive industries (e.g., household cleaning products or personal care) can innovate their product formulations and packaging to appeal to "Water Savers." Discuss specific examples from the report (e.g., solid formats, concentrated formulas, waterless products) and how these innovations can be marketed to highlight both environmental benefits and potential cost savings for consumers.

Acknowledgments ChatGPT, an AI language model developed by OpenAI, and Gemini, an AI model developed by Google, were used in this chapter of this textbook to support the development of case studies and to assist in generating discussion questions. The author has critically reviewed, verified, and adapted the outputs to ensure their accuracy, relevance, and alignment with the academic objectives of the work.

References

Becerril, J., & MacGuill, S. (2023, January). *Sustainability in tobacco.* Euromonitor International. https://www-portal-euromonitor-com.scd-rproxy.u-strasbg.fr/analysis/tab

Milasevic, M. (2024, January). *Retailer corporate strategies in sustainability.* Euromonitor International. https://www-portal-euromonitor-com.scd-rproxy.u-strasbg.fr/analysis/tab

Rajpara, A. (2024, October). *Sustainable fragrance claims: Emerging trends and prospects.* Euromonitor International. https://www-portal-euromonitor-com.scd-rproxy.u-strasbg.fr/analysis/tab

Tekutyte, E. (2024a, October 16). *Top five innovations in sustainability in Q2 2024.* Euromonitor International. https://www-portal-euromonitor-com.scd-rproxy.u-strasbg.fr/analysis/tab

Tekutyte, E. (2024b, November 29). *Top five innovations in sustainability in Q3 2024.* Euromonitor International. https://www-portal-euromonitor-com.scd-rproxy.u-strasbg.fr/analysis/tab

Tekutyte, E. (2025, February 3). *Top five innovations in sustainability in Q4 2024.* Euromonitor International. https://www-portal-euromonitor-com.scd-rproxy.u-strasbg.fr/analysis/tab

Zuniga, J. (2024, December). *Voice of the consumer: Sustainability survey 2024 key insights.* Euromonitor International. https://www-portal-euromonitor-com.scd-rproxy.u-strasbg.fr/analysis/tab

Zuniga, J. (2025, January 15). *Sustainability is a baseline for innovation.* Euromonitor International. https://www-portal-euromonitor-com.scd-rproxy.u-strasbg.fr/analysis/tab

Zuniga, J., & Tekutyte, E. (2024, December). *Innovation in sustainability.* Euromonitor International. https://www-portal-euromonitor-com.scd-rproxy.u-strasbg.fr/analysis/tab

Appendix

Appendix 1: Measures of Sustainable Value Delivery Through Behavioral Loyalty

Appendix 1.1: RFM Method for Assessing the Behavioral Loyalty of Sustainable Customers

D. K. Stoyanov, *Sustainable Marketing*, Springer Texts in Business and Economics,
https://doi.org/10.1007/978-3-032-16498-8

Step 1: Export the relevant data from your CRM software on the recency, frequency, and monetary value of purchases made by your sustainable customers.

Customer	Recency	Frequency	Monetary value
A	30 days	4 transactions	750.00 €
B	15 days	2 transactions	250.00 €
C	25 days	3 transactions	120.00 €
D	47 days	1 transactions	600.00 €
E	18 days	2 transactions	500.00 €
F	15 days	5 transactions	800.00 €
G	45 days	2 transactions	350.00 €
H	120 days	6 transactions	450.00 €
I	55 days	2 transactions	480.00 €
J	75 days	2 transactions	380.00 €
K	35 days	1 transactions	550.00 €
L	40 days	2 transactions	560.00 €
M	48 days	4 transactions	480.00 €
N	57 days	3 transactions	750.00 €
O	85 days	6 transactions	850.00 €
P	90 days	4 transactions	1,200.00 €

How by having all this information we can decide which are the most and least valuable customers in our database?

Custosmer	Let's assume that a company's customer database/portfolio composes of 16 clients (i.e., A, B, C, D, E, F, G, H, I, J, K, L, M, N, O, P)
Recency	Time elapsed since the customer's most recent purchase of a sustainable product (e.g., in days or months). "How recently did the customer engage with sustainable products or initiatives offered by the company?"
Frequency	Total number of sustainable product purchases within a defined time period (e.g., the last 365 days). "How frequently did the customer purchase sustainable products from the company in the past 12 months?"
Monetary value	Total amount of money spent on sustainable products over a specific time period (e.g., the last 365 days)."How much did the customer spend on the company's sustainable products in the past 12 months?"

Step 2: Analyze the distribution of each metric and segment customers by grouping values into bins.

$$Bw = \frac{Xmax - Xmin}{k}$$

Freedman-Diaconis rule: a method for selecting the bin width when constructing histograms that does not assume the data are normally distributed.

Bw *– bin width (size of the interval)*

Xmax *– maximum value in the dataset;*
Xmin *– minimum value in the dataset*
k *– number of categories in which we want to classify our customers*

Recency of purchases

$$Bw(R) = \frac{Xmax - Xmin}{k} = \frac{120 - 15}{3} = \mathbf{35}$$

Group 1 (Recency) = 35 + 15 = 50 => from 15 to 50

bin width + start/min value

Group 2 (Recency) = 35 + 50 = 85 => from 51 to 85

bin width + previous value

Group 3 (Recency) = 35 + 85 = 120 => from 86 to 120

bin width + previous value

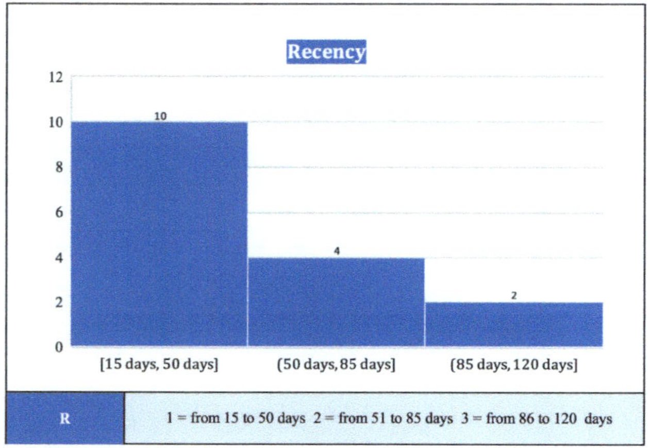

$$Bw = \frac{Xmax - Xmin}{k}$$

***Freedman-Diaconis rule*:** a method for selecting the bin width when constructing histograms that does not assume the data are normally distributed.

Bw – *bin width (size of the interval)*

Xmax – *maximum value in the dataset;*
Xmin – *minimum value in the dataset*
k – *number of categories in which we want to classify our customers*

Frequency of purchases

$$Bw(F) = \frac{Xmax - Xmin}{k} = \frac{6-1}{3} = 1{,}67 \approx \mathbf{2}$$

Group 1 (Frequency) = 2 + 1 = 3 => from 1 to 3

bin width + start/min value

Group 2 (Frequency) = 2 + 3 = 5 => from 4 to 5

bin width + previous value

Group 3 (Frequency) = 2 + 5 = 120 => from 6 to 7

bin width + previous value

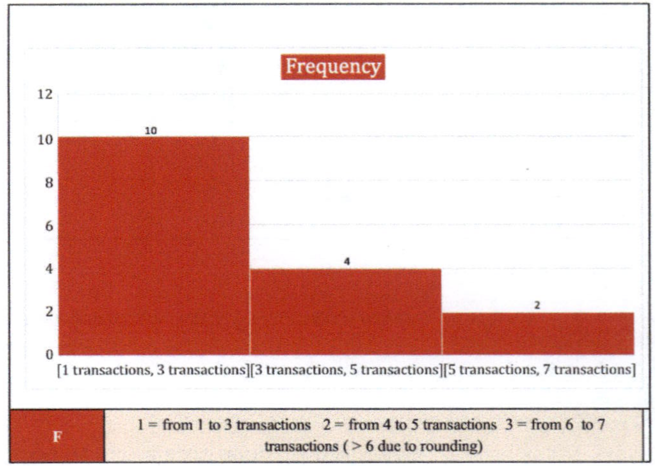

$$Bw = \frac{Xmax - Xmin}{k}$$

Freedman-Diaconis rule: a method for selecting the bin width when constructing histograms that does not assume the data are normally distributed.

$\boldsymbol{Bw}$ – *bin width (size of the interval)*

$\boldsymbol{Xmax}$ – *maximum value in the dataset;*
$\boldsymbol{Xmin}$ – *minimum value in the dataset*
$\boldsymbol{k}$ – *number of categories in which we want to classify our customers*

Monetary value of purchases

$$Bw(M) = \frac{Xmax - Xmin}{k} = \frac{1200-120}{3} = \mathbf{360}$$

Group 1 (Monetary value) = 360 + 120 = 480 => from 120 to 480

Group 2 (Monetary value) = 360 + 480 = 840 => from 481 to 840

Group 3 (Monetary value) = 360 + 840 = 1200 => from 841 to 1200

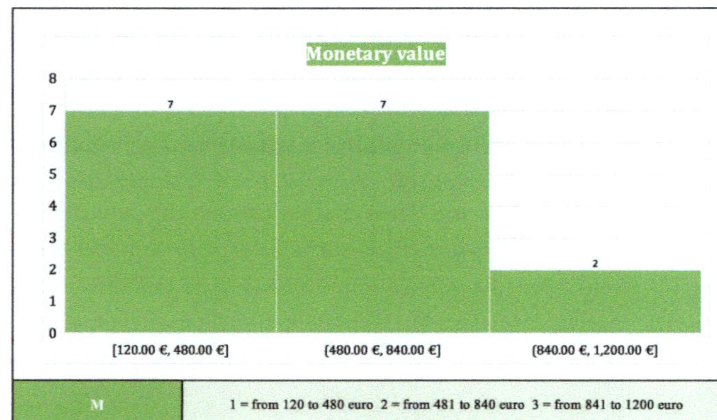

Bonus tip: *Alternative measures*

$$k = 1 + 3,322 Log_{10} n$$

Sturges' formula is used to objectively determine the number of categories (bins) into which a sample or total number of observations (n) can be divided. However, the number of categories can also be chosen subjectively, for example, based on the number of customer segments you want to serve differently, the size of the company, or available resources. The statistical assumptions for applying Sturges' formula are: 1) The observations in the dataset should be approximately normally distributed or at least not significantly skewed. 2) The sample size should be moderate, typically between 30 and 200 observations ($30 < n < 200$).

$$Bw = \frac{3,5\sigma}{\sqrt[3]{n}}$$

Scott's normal reference rule is used in some Excel versions to determine optimal histogram bin width by grouping observations (n). It works best for random samples from normally distributed data and uses the dataset's standard deviation (σ) to calculate bin size.

Step 3: Based on the previous stage add segment numbers to the RFM table: assign ranks to the R, F, M metrics

Custosmer	Recency	Frequency	Monetary value	R	F	M	RFMcode
A	30 days	4 transactions	750.00 €	1	2	2	122
B	15 days	2 transactions	250.00 €	1	1	1	111
C	25 days	3 transactions	120.00 €	1	1	1	111
D	47 days	1 transactions	600.00 €	1	1	2	112
E	18 days	2 transactions	500.00 €	1	1	2	112
F	15 days	5 transactions	800.00 €	1	2	2	122
G	45 days	2 transactions	350.00 €	1	1	1	111
H	120 days	6 transactions	450.00 €	3	3	1	331
I	55 days	2 transactions	480.00 €	2	1	1	211
J	75 days	2 transactions	380.00 €	2	1	1	211
K	35 days	1 transactions	550.00 €	1	1	2	112
L	40 days	2 transactions	560.00 €	1	1	2	112
M	48 days	4 transactions	480.00 €	1	2	1	121
N	57 days	3 transactions	750.00 €	2	1	2	212
O	85 days	6 transactions	850.00 €	2	3	3	233
P	90 days	4 transactions	1,200.00 €	3	2	3	323

Available customer segments on the market		
111	112	113
121	122	123
131	132	133
211	212	213
221	222	223
231	232	233
311	312	313
321	322	323
331	332	333

Step 4: Classify market segments into categories based on Recency of purchase (R).

R	F	M			Type of loyalty customers	Appropriate marketing strategy
1	1	1	2	3	**Active customers**	*Cross-selling, up-selling*
	2	1	2	3		
	3	1	2	3		
2	1	1	2	3	**Customers at risk**	*Retention campaigns*
	2	1	2	3		
	3	1	2	3		
3	1	1	2	3	**Customers that switched**	*Win back customers or end the relationship*
	2	1	2	3		
	3	1	2	3		

When will the Monetary value be more important? (M)
In a consumer durable products and most of the expensive and complex B2B products, the monetary value per transaction is normally high but frequency and recency is low. For example, you can't expect a customer to purchase a refrigerator or air conditioner on a monthly basis. In this case, a marketer could give more weight to monetary and recency aspects rather than the frequency aspect.

When will the Frequency be more important? (F)
When the purchasing cycle/frequency for a product is very high. For example, in a retail business selling fashion/cosmetics, a customer who searches and purchases products every month will have a higher recency and frequency score than monetary score. Accordingly, the RFM score could be calculated by giving more weight to R and F scores than M.

Custosmer	Recency	Frequency	Monetary value	R	F	M	RFMcode (text)	Type of customers	Appropriate marketing strategy
A	30 days	4 transactions	750.00 €	1	2	2	122	Active customers	Cross-selling, up-selling
B	15 days	2 transactions	250.00 €	1	1	1	111	Active customers	Cross-selling, up-selling
C	25 days	3 transactions	120.00 €	1	1	1	111	Active customers	Cross-selling, up-selling
D	47 days	1 transactions	600.00 €	1	1	2	112	Active customers	Cross-selling, up-selling
E	18 days	2 transactions	500.00 €	1	1	2	112	Active customers	Cross-selling, up-selling
F	15 days	5 transactions	800.00 €	1	2	2	122	Active customers	Cross-selling, up-selling
G	45 days	2 transactions	350.00 €	1	1	1	111	Active customers	Cross-selling, up-selling
H	120 days	6 transactions	450.00 €	3	3	1	331	Customers that switched	Win back customers or end the relationship
I	55 days	2 transactions	480.00 €	2	1	1	211	Customers at risk	Retention campaigns
J	75 days	2 transactions	380.00 €	2	1	1	211	Customers at risk	Retention campaigns
K	35 days	1 transactions	550.00 €	1	1	2	112	Active customers	Cross-selling, up-selling
L	40 days	2 transactions	560.00 €	1	1	2	112	Active customers	Cross-selling, up-selling
M	48 days	4 transactions	480.00 €	1	2	1	121	Active customers	Cross-selling, up-selling
N	57 days	3 transactions	750.00 €	2	1	2	212	Customers at risk	Retention campaigns
O	85 days	6 transactions	850.00 €	2	3	3	233	Customers at risk	Retention campaigns
P	90 days	4 transactions	1,200.00 €	3	2	3	323	Customers that switched	Win back customers or end the relationship

Step 5. Estimate the value of market segments

R	F	M		
		1	2	3
1	1	Bronze	Bronze	Silver
	2	Bronze	Silver	Gold
	3	Silver	Gold	Gold

Custosmer	Recency	Frequency	Monetary value	R	F	M	RFM code
A	30 days	4 transactions	750.00 €	1	2	2	122
B	15 days	2 transactions	250.00 €	1	1	1	111
C	25 days	3 transactions	120.00 €	1	1	1	111
D	47 days	1 transactions	600.00 €	1	1	2	112
E	18 days	2 transactions	500.00 €	1	1	2	112
F	15 days	5 transactions	800.00 €	1	2	2	122
G	45 days	2 transactions	350.00 €	1	1	1	111
H	120 days	6 transactions	450.00 €	3	3	1	331
I	55 days	2 transactions	480.00 €	2	1	1	211
J	75 days	2 transactions	380.00 €	2	1	1	211
K	35 days	1 transactions	550.00 €	1	1	2	112
L	40 days	2 transactions	560.00 €	1	1	2	112
M	48 days	4 transactions	480.00 €	1	2	1	121
N	57 days	3 transactions	750.00 €	2	1	2	212
O	85 days	6 transactions	850.00 €	2	3	3	233
P	90 days	4 transactions	1,200.00 €	3	2	3	323

Conclusion: Out of the most vauable market segment the company serves only Bronze and Silver consumers. It needs to develop further its marketing strategy to reach Golden consumers. Moreover. It needs to study further the customers at risk and those thath have potentially swtched to understand what kind of stratgies will be relevant to them.

Appendix 1.2: Assessing Behavioral Loyalty Through Retention Rate, Customer Lifetime Value, and Customer Equity and Share of Wallet in a Sustainability Context

$$r_t = \frac{(CE - CN)}{CS} \text{ x } 100$$

r_t – retention rate: the probability of a repeat purchase at time t. It is a measure of behavioral loyalty, calculated based on the company's overall customer database rather than on an individual customer.

CE - number of sustainable customers at end of period

CN - number of new sustainable customers acquired during period

CS - number of sustainable customers at start of period

$$CLV(s) = \sum_{t=0}^{n} \frac{(R_t - C_t) * r_t}{(1+i)^t} - AC \quad or \quad CLV(s) = \sum_{t=0}^{n} \frac{P_t * r_t}{(1+i)^t} - \text{AC}$$

$CLV(s)$ – sustainable customer life-time value

R_t – total revenue generated from the sustainable customer in period t

C_t – costs directly associated with serving the sustainable customer in period t

n – time horizon for estimating CLV (number of periods)

P_t – profit from the sustainable customer in t period

t – time period (e.g., year)

AC – acquisition costs (e.g., initial investment made to acquire the sustainable customer).

i – discount rate: the rate used to convert future profits into their present value, acknowledging that future earnings are less valuable than current ones. It typically ranges from around 5% in stable and predictable marketing environments to as high as 20% in highly dynamic and uncertain markets.

$$CE(s) = \sum_{i=1}^{n} CLV$$

$CE(s)$ – customer equity: the total combined sustainable customer lifetime values (CLVs) of all current and potential customers; in other words, the sum of all individual CLVs.

$$SOW(s) = \frac{\text{Customer's spending on brand/company}}{\text{Customer's total spending in the sustainable category}} \text{ x } 100$$

SOW(s) - sustainable share of wallet; the percentage of a customer's total spending within a particular sustainable product or service category that is allocated to a specific brand or company.

Business problem: "EcoTech Solutions" is a European B2B company that designs and supplies energy-efficient electronic components to industrial clients. The firm's mission is to promote sustainable innovation and help its customers reduce carbon emissions through green technologies. In 2020, EcoTech Solutions began the year with 12 international customers in its database. However, due to the market disruptions caused by COVID-19, it lost 4 customers by the end of the year. During the same period, the company acquired 2 new customers after participating in an industrial sustainability fair, which cost 250 €. One of these new clients was a large retailer called GreenBuild Retail Group (GBR)—a company committed to sourcing environmentally responsible products. EcoTech Solutions received an official confirmation that GBR will purchase energy-efficient products worth 600 € in 2021, with estimated servicing costs of 250 € for that year. The marketing manager of EcoTech Solutions wanted to ensure that the relationship with GBR would be mutually beneficial, long-term, and aligned with the company's sustainability values. Therefore, the manager forecasted GBR's Customer Lifetime Value (CLV) using the following projections.

Years	Expected Revenues (€)	Expected Costs (€)
2022	750 €	300
2023	800 €	325
2024	890 €	350
2025	920 €	375
2026	1000 €	400
2027	1200 €	425
2028	1300 €	450
2029	1450 €	475

1. Make a justified assumption about the discount rate based on the available information. You may consider market uncertainty, the firm's sustainability-driven risk tolerance, or its investment horizon. Then, calculate the sustainable customer lifetime value of GBR customer.
2. Estimate the company's sustainable customer equity.

Customer	CLVs (€)
A	1200
B	2605
C	600
D	1250
E	950
F	1650
G	1850

(continued)

(continued)

Customer	CLVs (€)
H	2981
I	2080
J = GBR client	To be calculated above

3. Assume that GBR's total annual spending on sustainable electronic components is 2400 €, of which EcoTech Solutions currently captures 600 €.

Business Solution:

Step 1. *Calculate the retention rate based on the company's overall customer database*

EcoTech Solutions began the year 2020 with 12 international customers and ended the year with 10 (after losing 4 but acquiring 2 new customers). Therefore, the retention rate is calculated as follows:

$$\boldsymbol{rt} = \frac{(10-2)}{12} \times 100 = 66.7\%$$

We apply this retention rate when estimating the CLV of the GBR client, by gradually adjusting the probability of retention over the forecasted relationship period. For example:

- In 2021, the first year of GBR's relationship, the retention probability is 100%, since the customer has just been acquired and makes their first purchase.
- In 2022, the retention probability decreases to 66.7%, based on the company's overall retention rate.
- In 2023, the probability of GBR remaining a customer becomes 66.7% × 66.7% = 44.5%.
- In 2024, it further decreases to 66.7% × 66.7% × 66.7% = 29.7%,
- and so on for each subsequent year.

This stepwise adjustment reflects the compounded probability that the customer will continue the relationship over time, based on EcoTech Solutions' historical ability to retain clients under uncertain market conditions.

Step 2. *Make an educated assumption on the discount rate*

Although there are various financial formulas specifically designed to estimate the discount rate, we do not want to deviate from the main focus of this example, which is on value delivery and behavioral loyalty measures. Therefore, we make the subjective qualitative assumption that the discount rate is 8%**.** Based on the available information, the environment in which EcoTech Solutions operates is relatively unstable and unpredictable, which justifies selecting a relatively low discount rate in this specific business case. To calculate the discounting factor for each year after the GBR customer made its first purchase (in 2021), we apply the following formula $(1+i)^t$, where i is the discount rate (0.08) and t represents the number of years since the initial purchase. For example:

- For the first year (2021): $(1+i)^1 = (1.08)^1 = 1.08$;
- For the second year (2022): $(1+i)^2 = (1.08)^2 = 1.17$
- and so on for each subsequent year.

Step 3. *Calculate the acquisition costs for the GBR client*

EcoTech Solutions initially invested €250 to participate in a sustainability fair, which resulted in acquiring two new customers, one of whom was GBR. Therefore, the customer acquisition cost for GBR represents half of the total investment, or €125.

Step 4 *Present a step-by-step estimation of sustainable customer lifetime value for EcoTech Solutions (see the table on the next page).*

Step 5. *Calculate sustainable customer equity as the sum of all CLVs, including that of the newly acquired GBR client.*

$$\begin{aligned} \text{CE} = {} & 1200 + 2605 + 600 + 1250 + 950 + 1650 \\ & + 1850 + 2981 + 2080 + 1657 = 16823€ \end{aligned}$$

This measure reflects the total long-term value of the company's sustainable customer base.

Step 6. *Estimate the sustainable share of wallet (SOWs) of the GBR client*

$$\text{SOW(s)} = \frac{600}{2400} \times 100 = 25\%$$

This measure reflects the percentage of the GBR customer's total spending on sustainable electronic components that is allocated to purchases from EcoTech Solutions. It indicates the extent to which GBR values and prioritizes its relationship with EcoTech Solutions within its overall sustainability-oriented procurement.

References

Customer Thermometer. (n.d.). *Customer retention rate.* https://www.customerthermometer.com/customer-retention/customer-retention-rate/

Fripp, G. (n.d.). *Videos CLV*. CLV Calculator. https://www.clv-calculator.com/videos-clv/

$$CLV(s) = \sum_{t=0}^{n} \frac{(R_t - C_t) * r_t}{(1+i)^t} - AC \quad or \quad CLV(s) = \sum_{t=0}^{n} \frac{P_t * r_t}{(1+i)^t} - \mathbf{AC}$$

Parameters	Time horizon (Years)									
	2020	2021	2022	2023	2024	2025	2026	2027	2028	2029
AC	125 €									
R_t		600 €	750 €	800 €	890 €	920 €	1,000 €	1,200 €	1,300 €	1,450 €
C_t		250 €	300 €	325 €	350 €	375 €	400 €	425 €	450 €	475 €
$P_t = (R_t - C_t)$		350 €	450 €	475 €	540 €	545 €	600 €	775 €	850 €	975 €
r_t		100.0%	66.7%	44.5%	29.7%	19.8%	13.2%	8.8%	5.9%	3.9%
$P_t * r_t$		350 €	300 €	211 €	160 €	108 €	79 €	68 €	50 €	38 €
$(1+i)^t$	1	1.08	1.17	1.26	1.36	1.47	1.59	1.71	1.85	2.00
$\frac{P_t * r_t}{(1+i)^t}$		378 €	350 €	266 €	218 €	158 €	126 €	117 €	92 €	76 €
$\sum_{t=0}^{n} \frac{P_t * r_t}{(1+i)^t}$		1,782 €								
$CLV(s) = \sum_{t=0}^{n} \frac{P_t * r_t}{(1+i)^t} - AC$	**1,657 €**									

MIX
Papier aus verantwortungsvollen Quellen
Paper from responsible sources
FSC® C105338

If you have any concerns about our products,
you can contact us on
ProductSafety@springernature.com

In case Publisher is established outside the EU,
the EU authorized representative is:
Springer Nature Customer Service Center GmbH
Europaplatz 3, 69115 Heidelberg, Germany

Printed by Libri Plureos GmbH
in Hamburg, Germany